Service Oriented Infrastructures and Cloud Service Platforms for the Enterprise

Theo Dimitrakos · Josep Martrat · Stefan Wesner
Editors

Service Oriented Infrastructures and Cloud Service Platforms for the Enterprise

A Selection of Common Capabilities Validated in Real-Life Business Trials by the BEinGRID Consortium

Editors

Theo Dimitrakos
Centre for Information and Security Systems
Research
BT Innovate & Design
PP13D–Orion Building
Adastral Park, Martlesham Heath
Suffolk IP5 3RE
UK
theo.dimitrakos@bt.com

Stefan Wesner
High Perfomance Computing Center
(HLRS)
University of Stuttgart
Nobelstrasse 19
70569 Stuttgart
Germany
wesner@hlrs.de

Josep Martrat
ATOS ORIGIN–Research and Innovation
Avda. Diagonal 200
08018 Barcelona
Spain
josep.martrat@atosresearch.eu

Front cover figure: The BEinGrid logo is copyrighted by CETIC and other BEinGrid consortium members. BEinGrid is partially funded by the European Comission under the contract number 034702.

ISBN 978-3-642-04085-6   e-ISBN 978-3-642-04086-3
DOI 10.1007/978-3-642-04086-3
Springer Heidelberg Dordrecht London New York

Library of Congress Control Number: 2009939137

Mathematics Subject Classification (2000): 68M14, 68U35

Cover design: WMXDesign GmbH, Heidelberg

Printed on acid-free paper

Springer is part of Springer Science+Business Media (www.springer.com)

Preface

The way enterprises conduct business today is changing rapidly and significantly. The enterprise has become more pervasive with a mobile workforce, distributed sites and outsourced data centers and is considering the use of even more cost-efficient computing platforms. These may run internally, or be offered by trusted providers and utilized only by selected business partners or may be open and run on public infrastructure. In addition, companies seeking to optimize their processes across their supply chains are implementing integration strategies that include their customers and suppliers rather than looking inward.

Today enterprises are enduring a strong pressure on cost reductions and an intensifying market competition forcing them to be more efficient, productive, agile and innovative in order to meet business objectives. Consequently there is increasing demand for technologies that help enterprises increase their customer base while reducing their costs and extending their competitive advantage.

It is important for any enterprise to understand how its business has performed at any given time in the past, now, and in the future. Bringing value networks together in the so-called Virtual Organizations (VO) over a shared IT infrastructure brings very strong benefits in terms of cost reduction, increased agility and shorter cycle-time and time-to-market. However, the presence of multiple authorities and complex relationships regarding the ownership of resources and information in contexts that span across organizational borders, mean that different authorities must be able to define policies about entitlements, ICT resource utilization and access. ICT resource administrators and resources may not necessarily belong anymore to the same organization. It therefore becomes much harder for an enterprise to govern its collaboration with other enterprises in a safe and controlled way.

Enterprises need better mechanisms to control how trust is established between business partners, how identities and other security attributes are shared, how security policy is defined and enforced—especially when policy applies on users and resources that are not controlled by a single enterprise. Improvements are also required for sharing and federating information and data efficiently across the value chain while ensuring privacy, confidentiality of corporate information and compliance to data protection. There is also a need for well-orchestrated, end-to-end operations management that provides controlled visibility, governance of network and

IT state, flexible license management models, understanding of how operations perform against Service Level Agreements (SLA) and timely assessment of the impact of security policy violations and the availability of resources.

Technologies such as Service Oriented Architecture (SOA) based Web Services, Grid Computing and more recently Cloud Computing, which we classify under the general term Service Oriented Infrastructures (SOI), form the basis of the technology tool-box that organizations utilize in modern enterprises in order to face such challenges. As Cloud Computing matures we expect these strategies to increasingly involve outsourcing models that integrate in-house and in-cloud services or integrate services hosted in different Cloud Computing platforms.

However there are still disparities between the research and technological advancements of the last decade in SOI and its uptake by the market to the extent that technological innovation is applied to bring real improvements in everyday business. This is a gap that BEinGRID tackled. BEinGRID, Business Experiments in GRID, is the European Union's largest integrated project funded by the Information Society Technologies (IST) research, part of the European Union's sixth research Framework Programme (FP6). This consortium of 96 partners was led by project management team in Atos Origin and a technical director from BT. *This book presents the main technical results of the BEinGRID project.*

The strategic mission of BEinGRID was to understand the commercial requirements for SOI use and to apply of these technologies in commercial environment, involving software vendors, IT integrators, service providers and end-users. The project run 25 business pilots in diverse economic sectors such as entertainment, pharmaceutical, engineering modeling for ship building or aeronautics, finance, textile industry logistics, earth observation, etc that helped on one hand to understand and extract common requirements and on the other hand to validate the designed and developed solutions. Technological innovation in BEinGRID focused on areas where we witnessed either significant challenges that inhibit widespread commercialization or where the anticipated impact of the innovation (i.e. the "innovation dividend") is particularly high.

As part of implementing this mission teams of technology experts and business analysts embedded in the pilot projects have attempted to reduce the adoption barriers by eliciting common technical requirements that solve common business problems across these vertical markets, by defining innovative generic solutions, called *common capabilities*, that meet these requirements, by producing design patterns that explain how these solutions can be implemented over commonly used commercial and experimental platforms and by elaborating best-practice guidelines demonstrating how these solutions can be applied in exemplar business scenarios. These contributions can be classified in the following areas:

- *Virtual Organization Management* capabilities help businesses establish secure, accountable and efficient collaborations sharing services, resources and information. These include innovations that enable the secure federation of autonomous administrative domains, and the composition of services hosted by different enterprises or in-cloud platforms.

- *Trust & Security* capabilities address areas where a perceived or actual lack of security appears to inhabit commercial adoption of SOI. These include solution for brokering identities and entitlements across enterprises, managing access to shared resources, analyzing and reacting to security events in a distributed infrastructure, securing multi-tenancy hosting, and securing the management of in-cloud services and platforms. These innovations underpin capabilities offered in Virtual Organization Management and other categories.
- *License Management* capabilities are essential for enabling the adoption of "pay-per-use" and other emerging business models, and had so far been lacking in the majority of SOI technologies including Grid and Cloud computing.
- Innovations to improve the management of *Service Level Agreements* cover the whole range from improvements to open standard schemes for specifying and negotiating agreements to solutions to ensuring fine-grained monitoring of usage, performance and resource utilization.
- *Data Management* capabilities enable better storage, access, translation and integration of data. Innovations include capabilities for aggregating heterogeneous data sources in virtual data-stores and ensuring seamless access to heterogeneous geographically distributed data sources.
- Innovations in *Grid Portals* enable scalable solutions based on emerging Web 2.0 technologies that provide an intuitive and generic instrumentation layer for managing user communities, complex processes and data in SOI.

The originality of BEinGRID findings and proposed solutions is that they have already been tried out. The results are not about theories or frameworks, but about real, tested, experimented, adapted solutions and the experiences gained by their use. The case studies that BEinGRID produced are real, conducted by a broad spectrum of European businesses which operate in the real world. Return-on-Investment has been examined, the legal context has been worked out, the technical problems have found a solution often involving a high innovation dividend.

We hope that this book will deliver the essential technical concepts of many years of research and innovation on Service Oriented Infrastructures in Europe in a condensed way and ultimately becomes a source of references for researchers and practitioners alike. The timing of this book is in line with the maturation of Grid computing and the emergence of Cloud Computing as the attraction of research interest in Service Oriented Infrastructures. The book offers a set of concepts and tools that will help companies in Europe and world-wide to adopt SOI technologies and to realize this transition successfully.

We would like to acknowledge the support from the European Commission for the BEinGRID project and prior research that provided the foundation for it. In particular we would like to thank the BEinGRID Project Officers Annalisa Bogliolo and Maria Tsakali for their continuous support in implementing this large project as well as the head of units Jesús Villasante and Wolfgang Boch for sharing with us the vision that research in Service Oriented Infrastructures and their embodiment as Grid or Cloud Computing is important for Europe.

London, Barcelona, Stuttgart,   Theo Dimitrakos
July 2009   Josep Martrat
                                                                                    Stefan Wesner

Contents

Contributors

George Beckett EPCC, The University of Edinburgh, James Clerk Maxwell Building, Mayfield Road, Edinburgh EH9 3JZ, UK, g.beckett@epcc.ed.ac.uk

David Brossard Centre for Information Systems and Security, BT Innovate & Design, PP13D, Ground Floor, Orion Building, Adastral Park, Martlesham Heath IP5 3RE, UK, david.brossard@bt.com

Nicola Capuano Centro di Ricerca in Matematica Pura ed Applicata (CRMPA) c/o DIIMA, via Ponte Don Melillo, 84084 Fisciano (SA), Italy, capuano@crmpa.unisa.it

Antonio Conguista Centro di Ricerca in Matematica Pura ed Applicata (CRMPA) c/o DIIMA, via Ponte Don Melillo, 84084 Fisciano (SA), Italy, congiusta@crmpa.unisa.it

Theo Dimitrakos Centre for Information and Security Systems Research, BT Innovate & Design, PP13D, Ground Floor, Orion Building, Adastral Park, Martlesham Heath IP5 3RE, UK, theo.dimitrakos@bt.com

Ana Maria Juan Ferrer Atos Origin, calle Albaracín 25, 28037 Madrid, Spain, ana.juanf@atosorigin.com

Angelo Gaeta Centro di Ricerca in Matematica Pura ed Applicata (CRMPA) c/o DIIMA, via Ponte Don Melillo, 84084 Fisciano (SA), Italy, agaeta@crmpa.unisa.it

Matteo Gaeta Centro di Ricerca in Matematica Pura ed Applicata (CRMPA) c/o DIIMA, via Ponte Don Melillo, 84084 Fisciano (SA), Italy, gaeta@crmpa.unisa.it

André Gemünd Fraunhofer SCAI, Schloss Birlinghoven, 53754 Sankt Augustin, Germany, andre.gemuend@scai.fraunhofer.de

Piotr Grabowski Poznan Supercomputing and Networking Center, Noskowskiego 10, Poznan 61-704, Poland, piotrg@man.poznan.pl

Alex Gusmini Centre for Information Systems and Security, BT Innovate & Design, PP13D, Ground Floor, Orion Building, Adastral Park, Martlesham Heath IP5 3RE, UK, alex.gusmini@bt.com

Domenic Jenz HLRS, Nobelstr. 19, 70569 Stuttgart, Germany

Efstathios Karanastasis National Technical University of Athens, 9 Iroon Polytechniou, Zografou 15773, Greece, karanastasis@telecom.ntua.gr

Kostas Kavoussanakis EPCC, The University of Edinburgh, James Clerk Maxwell Building, Mayfield Road, Edinburgh EH9 3JZ, UK, k.kavoussanakis@epcc.ed.ac.uk

Ottmar Krämer-Fuhrmann Fraunhofer SCAI, Schloss Birlinghoven, 53754 Sankt Augustin, Germany

Roland Kuebert HLRS, Nobelstr. 19, 70569 Stuttgart, Germany, kuebert@hlrs.de

Stéphane Mouton SST-SOA, CETIC, Bâtiment Éole, Rue des Frères Wright, 29/3, 6041 Charleroi, Belgium, stephane.mouton@cetic.be

Francesco Orciuoli Centro di Ricerca in Matematica Pura ed Applicata (CRMPA) c/o DIIMA, via Ponte Don Melillo, 84084 Fisciano (SA), Italy, orciuoli@crmpa.unisa.it

Mark Parsons EPCC, The University of Edinburgh, James Clerk Maxwell Building, Mayfield Road, Edinburgh EH9 3JZ, UK, m.parsons@epcc.ed.ac.uk

Michal Piotrowski EPCC, The University of Edinburgh, James Clerk Maxwell Building, Mayfield Road, Edinburgh EH9 3JZ, UK, m.piotrowski@epcc.ed.ac.uk

Robert Piotter HLRS, Nobelstr. 19, 70569 Stuttgart, Germany, piotter@hlrs.de

Yona Raekow Fraunhofer SCAI, Schloss Birlinghoven, 53754 Sankt Augustin, Germany

Igor Rosenberg Atos Origin, calle Albaracín 25, 28037 Madrid, Spain, igor.rosenberg@atosorigin.com

Mark Sawyer EPCC, The University of Edinburgh, James Clerk Maxwell Building, Mayfield Road, Edinburgh EH9 3JZ, UK, m.sawyer@epcc.ed.ac.uk

Horst Schwichtenberg Fraunhofer SCAI, Schloss Birlinghoven, 53754 Sankt Augustin, Germany, horst.schwichtenberg@scai.fraunhofer.de

Christian Simmendinger T-Systems SFR, Pfaffenwaldring 38-40, 70569 Stuttgart, Germany

Craig Thomson EPCC, The University of Edinburgh, James Clerk Maxwell Building, Mayfield Road, Edinburgh EH9 3JZ, UK, c.thomson@epcc.ed.ac.uk

Arthur Trew EPCC, The University of Edinburgh, James Clerk Maxwell Building, Mayfield Road, Edinburgh EH9 3JZ, UK, a.trew@epcc.ed.ac.uk

Theodora Varvarigou National Technical University of Athens, 9 Iroon Polytechniou, Zografou 15773, Greece, dora@telecom.ntua.gr

Chapter 1
Introduction

Theo Dimitrakos

Abstract Service Oriented Infrastructures including Grid and Cloud Computing are technologies in a critical transition to wider adoption by business. Their use may enable enterprises to achieve optimal IT utilization, including sharing resources and services across-enterprises and on-demand utilization of those made available by business partners over the network. This chapter presents an introduction to, and an overview of, a selection of common capabilities (i.e. services capturing reusable functionality of IT solutions) that have been applied to tackle challenging business problems and were validated in real-life business trials covering most European market sectors. The remaining of the book elaborates these results, explains the process that was used to produce them.

1.1 Motivation

The way enterprises conduct business today is changing greatly. The enterprise has become more pervasive with a mobile workforce, distributed sites, outsourced data centers and considering the use of even more cost-efficient Cloud Computing platforms [32]. These may run internally (so called "internal Clouds"), or be offered by trusted providers and utilized only by selected business partners (so called "partner Clouds") or may be open and run on public infrastructure (so called "public Clouds"). In addition, companies seeking to optimize their processes across their supply chains are implementing integration strategies that include their customers and suppliers rather than looking inward. This increases the need for managing business-to-business (B2B) and business-to-business-to-customer (B2B2C) collaborations and securing end-to-end transactions between business partners and the customer. As Cloud computing matures we expect these strategies to increasingly involve outsourcing models that integrate in-house and in-cloud services, thus yielding "hybrid" Cloud platforms [31], or integrate services hosted in different cloud platforms, thus yielding "federated" Cloud platforms.

Business conduct has always been information centric—the combination of fast discovery of, and access to, reliable information and the ability to process this information and act swiftly upon it, is a common characteristic of successful businesses in the 21st century. Hence it is very important for corporate IT to provide the

T. Dimitrakos (✉)
Centre for Information and Security Systems Research, BT Innovate & Design, PP13D, Ground Floor, Orion Building, Adastral Park, Martlesham Heath IP5 3RE, UK
e-mail: theo.dimitrakos@bt.com

T. Dimitrakos et al. (eds.), *Service Oriented Infrastructures and Cloud Service Platforms for the Enterprise*, DOI 10.1007/978-3-642-04086-3_1, © Springer-Verlag Berlin Heidelberg 2010

means for sharing and federating information and data efficiently across the value chain in order to meet business objectives. This becomes even more challenging as businesses generate vast amounts of information and data at different levels of sensitivity including customer data that are subject to different legislation about its storage and transfer depending on where business operates and where the customer is based.

Globalization and agility of integration require more systems along with more partners and more constraints and produce more complex environments where decision making processes are equally increasingly complex and crucial for this connected organization. Change in a single process has the potential to impact more than one partner and disrupt a wider range of business processes.

It is important for any enterprise to understand how its business performs at any given time in the past, now, and in the future. However, single partners no longer have a full visibility of all processes and their consequences. It becomes much harder for a single enterprise to therefore govern its collaboration with other enterprises in a safe and controlled way, to understand the use of its information and resources across the value chain, and to identify and assess the impact of violations of policies or agreements. There is a need for well-orchestrated, end-to-end Operations management that provides controlled visibility, governance of network and IT state, flexible license management models, understanding of how operations perform against Service Level Agreements (SLA) and timely assessment of the impact of security and IT resource usage policy violations and of the availability of resources. There is consequently an increasing need for end-to-end operation dashboards showing real-time state of the corporate infrastructure, including the B2B integration points, in relation to Key Performance Indicators that reflect business objectives and their priorities.

As pervasive organizations connect their heterogeneous environments and systems, cross- and intra-enterprise compliance becomes more critical. The legal and regulatory frameworks become more complex and less forgiving [17]. Companies have to comply with their own directives and regulations as well as with different legislations and regulations depending on the region of operation and the client or partner organizations' rules and legal constraints. IT use in the corporate environment, and in particular the governance of the IT infrastructure that enables business services, will need to provide means to measure and control such complex compliance scenarios.

The presence of multiple authorities and complex relationships regarding the ownership of resources and information across different business contexts that span across organizational borders mean that multiple administrators must be able to define policies about entitlements, resource usage and access. Resource administrators and resources may not necessarily belong anymore to the same organization. For the corporate IT infrastructure this underlines the need for handling how information and resource usage and access policies that originate from different stakeholders are enforced over a common shared infrastructure.

As the workforce becomes mobile, and the organizations increase their distribution and further integrate their collaborations and share their resources, the risks associated with the exposure of corporate information assets, services and resources increase. It becomes essential that, once threats are identified, a coordinated reaction is per-formed in real time to adapt usage and access policies as well as business process parameters across the value chain in order to mitigate the risks.

Finally, another consequence of these changes in the organizational environment is the emergence of the notion of Virtual Organizations (VO). These are defined in [14] as temporary or permanent coalitions of individuals, groups, organizational units or entire organizations that pool resources, capabilities and information to achieve common objectives. According to this definition, VOs can provide services and thus participate as a single entity in the formation of further VOs, hence creating recursive structures with multiple layers of "virtual" value-adding service providers. The required scalability, responsiveness, and adaptability, requires a cost effective trust and contract management solution for dynamic VO environments.

Technologies such as Service Oriented Architecture (SOA) based Web Services, Grid Computing, Utility Computing and more recently Cloud Computing form the basis of the technology tool-box that Chief Architects tend to utilize in modern enterprises in order to meet these challenges. The use of these technologies, which we classify under the wider term "Service Oriented Infrastructures" (SOI), has the potential to bring many benefits to the business. However Service Oriented Infrastructures (SOI) technologies—and especially Service-Oriented Grid and Cloud Computing—are in a critical transition from research and academic or experimental use to wider adoption by business.

Effective solutions addressing these challenges require interdisciplinary approaches integrating tools from information technology and computer science with tools from law, economics and business management. Furthermore, the lack of business reference cases to persuade customers to explore the economic benefits of SOI hampers commercial exploitation of this important new technology solution across the Europe. It is well known that the point of greatest peril in the development of a market for new technologies lies in making the transition from an early market dominated by a few visionary customers to a mainstream market dominated by a large community of customers who are predominantly pragmatists. Crossing this chasm must be the primary focus of any long-term high-technology marketing plan. A successful crossing is known to lead to success. At this critical transition, however, businesses are not aware of all the benefits and weaknesses from new technologies, they are often confused by the hype generated around emerging technologies and they don't have business cases to refer to. This lack of in-depth knowledge limits the success of commercially exploiting valid results and delays the maturation of a market, weakening competitiveness and leadership in this technological area.

Since 2003 the author has been the technical director of multidisciplinary research and innovation programmes[1] developing solutions for these challenges together with renowned academic researchers, leading infrastructure and service

---

[1]Examples of such projects include: BEinGRID (www.beingrid.eu)—the technical results of which project are the main focus of this chapter—TrustCoM (www.eu-trustcom.com) where the author led a consortium that brought together experts from academia, service providers such as BT, integrators such Atos Origin and IBM, product vendors such as Microsoft and SAP, as well as cus-

providers, product vendors, technology integrators and technology consumers. Over the years many of these results have been already transformed into solutions that enabled companies to transform their organization to a Service Oriented Enterprise. One of the most important, successful and rewarding projects has been the BEin-GRID project—Europe's largest and most encompassing research and innovation initiative on business applications of service oriented IT. The rest of this chapter summarizes how this initiative produced and validated innovations that help enterprises to meet the challenges mentioned above.

1.2 The BEinGRID Project

BEinGRID, Business Experiments in GRID, is the European Union's largest integrated project funded by the Information Society Technologies (IST) research, part of the European Union's sixth research Framework Programme (FP6). This consortium of 96 partners is drawn from across the EU and represents the leading European organizations in SOI and Grid Computing and a broad spectrum of companies covering most vertical markets keen on assessing the benefits to their productivity, competitiveness and profitability from using Grid and Cloud Computing solutions. The consortium is led by project management team in Atos Origin and a technical director from BT.

The mission of BEinGRID is to generate knowledge, technological improvements, business demonstrators and reference case-studies to help companies in Europe and world-wide to establish effective routes to foster the adoption of SOI technologies such as Grid and Cloud Computing technologies and to stimulate research that helps realizing innovative business models using these technologies. In terms of technology innovation BEinGRID has defined and steered the technical direction of Business Experiments (BE) in all vertical market sectors by offering them best-practice guidance in each of the stages (requirements, design, prototyping, demonstration), thought-leadership in tackling innovative problems and technical advice for improving the BE solution.

As part of implementing this mission teams of technology experts and business analysts by eliciting common technical requirements that solve common business problems across vertical markets, by defining innovative generic solutions, called *common capabilities*, that meet these requirements, by producing design patterns that explain how these solutions can be implemented over commonly used commercial and experimental platforms and by producing best-practice guidelines demonstrating how these solutions can be applied in exemplar business scenarios. The remaining of this chapter presents an introduction to the BEinGRID programme and a high-level overview the main technical results it produced.

---

tomers such as BAe Systems—and iTrust (www.itrust.uoc.gr) that was established in 2002 and has now grown in a multi-disciplinary community that brings together researchers and practitioners across most continents under the auspices of the IFIP working group on trust management (www.ifip.org).

1.2.1 The BEinGRID Matrix

To meet these objectives, BEinGRID has undertaken a series of targeted Business Experiments (BEs) designed to implement and deploy Grid solutions across a broad spectrum of European business sectors including the media, financial, logistics, manufacturing, retail, and textile sectors. The consortium conducted twenty-five pilot case studies that have been summarized at the BEinGRID project Web site [11] and presented in book [10]. Each one of these twenty-five BEs is a showcase of a real-life pilot application focusing on a specific business opportunity and addressing current customer needs and requirements. The involvement of all actors in a representative value chain including consumers and service providers has been considered crucial for producing successful case studies that build on the experiences of early adopters. Consequently participation of representative consumers and of providers that can take a solution to the market has been ensured in each of the BEs. The BEinGRID Business Experiments has been classified according to their main vertical market, the business model they exploit, and the technological innovations they validate.

1.2.1.1 Vertical Market Sectors

Each BE focuses on a particular vertical market and addresses concrete business issues and in which the main actors of the value network are represented. From this perspective, the 25 BEs of BEinGRID cover the following sectors:

- *Advanced Manufacturing*. This class comprises BE that apply Grid technology to design products or components that are later manufactured or to optimize some part of the production processes.
- *Telecommunications*. This sector covers the BE that use SOI in order to improve existing or offer new innovative services that can improve the operational cost and the quality of services offered by net-work operators. These include services for sharing data and services among network operators and detecting fraud when roaming.
- *Financial*. This sector includes the solutions used by financial organizations to optimize existing business activities or to produce new and innovative services to their customers.
- *Retail*. This sector includes BE that improve the business activities related to management of goods (acquisition, delivery, transformation . . .).
- *Media & Entertainment*. This sector consists of BE related to the management and processing of media content (capture, rendering, post-production, delivering) and, more broadly, the provision of on-line entertainment services including scalable and high-performing collaborative gaming.
- *Tourism*. This sector covers the BE that is used by the tourism industry in order to optimize existing business activities or to produce new and innovative services to their customers.
- *Health*. This sector covers the BE that focus on processing of medical data, compute intensive algorithms for medical science and provision of services that allow

optimize the cycle time, the quality and the cost of medical treatment covering all
actors contributing to the treatment.

- *Environmental Sciences* covers the BE that focus on processing geophysical data
  and apply compute intensive algorithms to science focusing on the analysis of
  and protection against damage to the Environment and natural disasters.

Different BE use different middleware in the same sector in order to solve specific real-world challenges. The anticipated commercial and social impact and innovation dividend have been the main criteria in selecting the BE, in addition to the
necessary use of Service Oriented Infrastructure technologies including Grid and
Cloud Computing.

1.2.1.2 Business Models

The business models explored in these pilot projects have been categorized based
on criteria that take into account their value propositions, their technological and
economic incentives and emerging trends in the market of SOI technologies (e.g.
Grid and Cloud Computing).

One category focuses on achieving optimizing cycle time and costs by improving
resource utilization. At the core of this category are innovations facilitating:

(a) better utilization of compute power and data storage,
(b) on-demand provision of additional compute power and storage in order to re-
    spond to peaks in consumption, and
(c) aggregation of heterogeneous data sources in virtual data-stores.

Another category focuses on collaboration and resource sharing. At the core of
this category are innovations improving:

(a) the agility of businesses and their ability to respond to business opportunity by
    enabling the swift establishment of multi-enterprise collaborations,
(b) the execution of collaborative processes spanning across-enterprise boundaries,
(c) provision of, and access to, shared network-hosted ("cloud") services that facil-
    itate collaboration, and
(d) seamless access to heterogeneous geographically distributed data sources.

Another family of categories focused on new service paradigms centered on
"pay-as-you-go" (PAYG) and new paradigms of ICT services (*-aaS) including
Software as a Service (SaaS), Platform as a Service (PaaS) and Infrastructure as
a Service (IaaS).

More information on the business analysis results is available in a book "Grid
and Cloud Computing: A Business Perspective on Technology and Applications"
by Katarina Stanoevska-Slabeva, Thomas Wozniak and Santi Ristol, and on the
Web [21].

1.2.1.3 Research and Technological Innovation Themes

The technological advancements and innovations inspired or validated by the BE
have been categorized in the following *thematic areas*. These are areas where we

witnessed either significant challenges that inhibit widespread commercialization or where the anticipated impact of the innovation (i.e. the "innovation dividend") is particularly high.

- *Virtual Organization Management* capabilities help businesses establish secure, accountable and efficient collaborations sharing services, resources and information. These include innovations that enable the secure federation of autonomous administrative domains, and the composition of services hosted by different enterprises or in-cloud platforms.
- *Trust & Security* capabilities address areas where a perceived or actual lack of security appears to inhabit commercial adoption of SOI. These include solution for brokering identities and entitlements across enterprises, managing access to shared resources, analyzing and reacting to security events in a distributed infrastructure, securing multi-tenancy hosting, and securing the management of in-cloud services and platforms. These innovations underpin capabilities offered in Virtual Organization Management and other categories.
- *License Management* capabilities are essential for enabling the adoption of PAYG and other emerging business models, and had so far been lacking in the majority of SOI technologies including Grid and Cloud computing.
- Innovations to improve the management of *Service Level Agreements* cover the whole range from improvements to open standard schemes for specifying agreements to solutions to ensuring fine-grained monitoring of usage, performance and resource utilization.
- *Data Management* capabilities enable better storage, access, translation and integration of data. Innovations include capabilities for aggregating heterogeneous data sources in virtual data-stores and ensuring seamless access to heterogeneous geographically distributed data sources.
- Innovations in *Grid Portals* enable scalable solutions based on emerging Web2.0 technologies that provide an intuitive and generic instrumentation layer for managing user communities, complex processes and data in SOI.

Technological innovation in BEinGRID focused on areas where we witnessed either significant challenges that inhibit widespread commercialization or where the anticipated impact of the innovation (i.e. the "innovation dividend") is particularly high. It built on the experience and research by teams of experts embedded in the pilot projects across vertical market sectors (BEinGRID Business Experiments) and used some of these pilot projects in order to demonstrate and validate the technology innovation in a business context. The technological innovation results take the form of core, generic functionality or processes that can be implemented over commercial and experimental service oriented middleware and infrastructures in order to add or help realize business value that is known to be important for commercial success. These technological innovation results have been delivered by means of the following outputs of the programme:

- *Common technical requirements* that identify specific challenges where technical innovation is required. These were *elicited* by analyzing BEs across vertical market sectors; their *interdependences* have been analyzed within and across thematic

areas; and they have been prioritized in terms of *innovation potential* and anticipated *business impact* based on feedback from BEs in several market sectors and
criticality[2] in terms of their interdependences.

- *Common capabilities* that capture the generic functionality that would need to be
  in place in order to address these requirements. These are necessary for enhancing
  current service offerings and delivery platforms in order to meet the business
  challenges described at the introduction of this chapter.
- *Design patterns* that describe one or more possible solutions that describe how
  systems may be architected in order to realize each common capability.
- *Reference implementations* that realize selected common capabilities over commercial middleware. These were subject to quality assurance processes including:
  release testing (focusing on robustness, installation and usability of artifacts);
  conformance testing to assure that the artifacts are adequately implementing the
  functionality of the capability; *documentation* and *training material* explaining
  how to deploy, integrate and improve the artifacts.
- *Integration scenarios* illustrating how a critical mass of interdependent common
  capabilities can be implemented together to maximize added value.
- *Validation scenarios* illustrating the benefits of implementing selected common
  capabilities to enhance business solutions in real-life case-studies.
- *Best-practice guidelines* explaining how these common capabilities can be taken
  advantage of in indicative business contexts.

1.2.1.4 Knowledge Repository for SOI and Cloud Computing

Research and innovation in the BEinGRID programme is complemented with the
development of a public knowledge and toolset repository [21] that aims to con-
centrate the biggest and most valued selection of service designs, best practices,
case studies, technology implementations, and other resources that may enable the
adoption of SOI technologies such as Grid and Cloud Computing. This knowledge
repository also includes descriptions of the capabilities produced by the BEinGRID
project, as well as information and/or software for their reference implementations
and other auxiliary content such as technical reports, white papers, presentations,
demonstration videos and training material.

1.3 Common Capabilities for SOI and Cloud Services

In the remaining of this chapter, we provide an overview of the innovative solutions
produced, of the research challenges that were addressed, the business drivers that
motivated the development of these solutions, and their anticipated business impact
(i.e. their "innovation dividend") based on the experience generated by the Business
Experiments where these results have been validated.

---

[2]In simple terms, criticality of a technical requirement is a function of the number and relative
priority of other requirements that depend upon it.

1.3.1 Life-Cycle Management of Virtual Organizations

Based on the analysis of B2B collaborations in the 25 business experiments of the BEinGRID project, the following two were identified as the most significant recurring issues during the B2B collaboration life-cycle [19]:

1. The identification and selection of business partners (based on their reputation and the suitability of services that they offer) among an available pool of service providers or consumers.
2. The creation and management of a Circle-of-Trust among the selected partners.

The "VO Set-up" common capability has been developed by the BEinGRID project to offer a standards-based foundation for business solutions to these problems. This capability facilitates the identification of partners in some B2B collaboration and the creation and lifecycle management of a circle of trust among business partners.

It is useful in typical B2B collaborative scenarios where participants (corporate users, services, resources) have to be identified and trust has to be established with each other. A demand for including new participants can appear during the collaboration lifetime, and the existing participants (or their organization) may be dropped. The security of the collaboration also needs to be maintained: the businesses participating in B2B collaborations must be able to identify one another, identify messages as coming from other members of the same B2B collaboration, and establish the validity of security claims made by other parties in the B2B collaboration about the identity and entitlements of a user or other resource.

A competitive differentiator with respect to alternative solutions is that trust is aligned to consumer/provider relationships, hence supporting the evolution of a Circle-of-Trust towards a trust network that reflects supply network relationships.

A high level architecture of this capability is shown in the next picture together with a summary of its functionality for each phase of a typical VO lifecycle and a possible deployment of its building blocks. To allow the secure federation lifecycle management, the VO Set Up interacts (via the federation manager building block) with the Security Token Service (STS) component developed in security theme of BEinGRID and presented in Sect. 1.3.2. In Fig. 1.1, the FM (Federation Manager) interface is a component offering a programmatic interface that allows the decoupling of the VO Set Up capability from the specific STS implementation thus enabling the VO Set-up capability to instrument heterogeneous STS implementations that agree on a basic Web-services interface pattern.

Each partner of a VO needs to be associated with some Security Token Service (STS), which acts as an identity broker enabling their participation in a Circle-of-Trust. The "VO Set Up" capability and its building blocks can be offered as in-cloud services or be deployed at the site of one of the business partners.[3]

---

[3]An analysis by BEinGRID indicated that most collaborators are willing to consider an in-cloud capability for this functionality. This preference is particularly high among companies that are

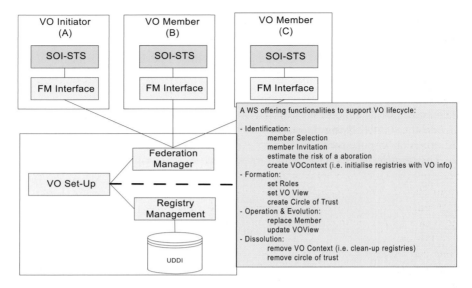

Fig. 1.1 The high-level architecture of the "VO Set-Up" common capability

This capability combines provider and service registries with identity federation management in a single loosely-coupled solution. Registries are built on top of UDDI standard [34] and allow the publication, discovery, and update of VO members and services. The secure federation model builds on the SAML [30] and WS-Federation [38] models and the results of the TrustCoM project [15]. The model is attribute and policy-based and allows the establishment of directed trust relationships that are associated with a common federation context. The following list summarizes some of the most significant improvements achieved by this architecture.

1. It can mange multiple, distinct and co-evolving B2B collaborations even if they include the common participants using common identity providers.
2. For each B2B collaboration context, the trust relationships between the (identity brokers of) business partners reflect the structure of the value network of this collaboration.
3. It enables evaluating the risk associated with a collaboration based on trust in each participant. In its current implementation, the risk is estimated by evaluating a weighted mean of "reliability" values associated to each member.[4]

---

used in participating in eCommerce hubs or similar. However, in scenarios where a main contractor is managing a B2B collaboration consisting of mainly subcontractors to this main contractor, a business partner appears to be equally popular or preferable for offering this capability.

[4]An analysis conducted by the BEinGRID project indicates that most collaborators are willing to consider an in-cloud capability for this functionality. This preference is particularly high among companies that are used in participating in eCommerce hubs or similar. However, in some situations deployment at a business partner appears to be equally popular or preferable: these are scenarios

The business benefits of this capability include offering instrumentation and co-ordination layers that act as 'glue' among different capabilities that are required during the life-cycle of B2B collaborations. Without the adoption of such a capability, providers willing to initiate or join in a B2B collaboration would need to deploy, manage and integrate a plethora of bespoke software components such as business registries, service registries, and Identity and Access Management solutions and implement a complex coordination process on top of them thus increasing cycle-time and cost and intensifying the risk of mistakes and failure due to incompatibility at the edges of bespoke solutions built to serve different objectives.

Early experimentation has indicated that the cycle-times of identification of partners and the establishment of a circle of trust among selected partners are reduced from 60% to 90% (depending on the investment already in place) with analogous cost reductions. Overall, the main benefit of this capability is that it offers organizations of all sizes the flexibility and dynamism they need in order to quickly exploit new business opportunities. This capability has been evaluated in the context of concrete case studies including a Virtual Hosting Environment for Distributed Online Gaming [3], supply chains in agriculture [6] and collaborative engineering. More information is provided in Chap. 3 of this book.

1.3.2 Federated Identity and Access Management

The need for security for agile business operations is so strong that, according to [17], despite the worldwide economic crisis—or possibly because of it—security aspects such as identity and access management (IAM) remain a critical undertaking for enterprises of all sizes and market sectors. Through increasing business-level visibility led by data-breach headlines, security spend continues to rise and take a growing share of overall IT spending. Indeed, IAM alone represents a growing market which accounted for almost $3 billion in revenue for 2006 [17]. According to Forrester, Market Overview (April 2009), security initiatives will focus on: (a) protecting data, (b) streamlining costly or manually intensive tasks, (c) providing security for an evolving IT infrastructure, and (d) understanding and properly managing IT risks within a more comprehensive enterprise framework.

In order to achieve agility of the enterprise and shorten concept-to-market timescales for new products and services, IT and communication service providers and their corporate customers alike increasingly interconnect applications and exchange data in a Services Oriented Architecture (SOA). Key security challenges come from this evolution of the way businesses interact nowadays as presented in the introduction of this chapter:

- A work environment that becomes pervasive with a mobile workforce;

---

where a main contractor is managing a B2B collaboration consisting of mainly subcontractors to the same main contractor.

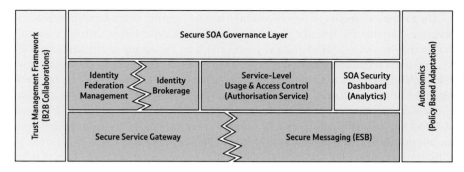

Fig. 1.2 Overview of the security capabilities required by service-oriented enterprises

- Outsourced Data Centers and in-cloud services;
- Business process integration with customers and suppliers across value chains;
- Many sources of identity and policy enforced over shared IT infrastructure;
- Manage access to resources in environments that are not under one's control;
- Ensure accountability over a mixed control infrastructure;
- Collect evidence about policy compliance for diverse regulatory frameworks.

For security to work, the mechanisms put in place must support, not hinder, such rich and flexible scenarios. They must be flexible and adaptive. In line with this analysis, security efficiency, with lower costs and improved service, security effectiveness, including regulatory compliance and business agility and increased productivity were the three main business drivers which influenced the activities of the Security area of the BEinGRID project [20]. In this section we focus only on two representative examples of the security capabilities: federated identity brokerage and distributed access management. For a full list of security capabilities, shown in Fig. 1.2, please refer to Chap. 4 of this book.

These security capabilities have been validated in a Business Experiment in eHealth [9], demonstrating the secure integration of a "cloud" HPC capability into a regional NHS network in Spain in order to facilitate the fast processing of radiotherapy analysis results while preserving the patient privacy and ensuring correctness of the association between patients and their radiotherapy examination data. All these three capabilities together with relevant capabilities from VO Management have also been validated in a trial demonstrating a network-centric distributed platform for scalable, collaborative on-line gaming [3].

1.3.2.1 Identity Brokerage and Identity Federation Context Management

This is a capability enabling identity federation and brokerage across business partners. Early developments of this capability stemmed from collaborative research between BT and the European Microsoft Innovation Center in the TrustCoM project [14]. It is a customizable platform for Identity-as-a-Service (IDaaS) provision with technological innovations that resulted in the following differentiators compared to what is currently available in the market:

The business logic of the identity broker adapts to optimize for each identity federation context. This innovation enables applying different authentication procedures, different federated identity standards, attribute types and entitlements on the same user or resource depending on the purpose of a B2B interaction and the scope of the identity federation. For each context the identity broker takes the embodiment of a behaviorally distinct instance of a Security Token Service (STS).

Administrators can author declarative policies to control information disclosure within the scope of each identity federation. Users can also author policies to control disclosure of user-provided data. In effect, different policies may apply on the same personal data used for different purposes in the same scope or used in different identity federations.

The identity broker has been designed with compliance in mind. An innovative policy issuance mechanism allows associating an administrator's identity with a digital signature of a policy fragment (or a user's identity with digital signatures of user-generated data) and providing evidence of use of policy and data in compliance with explicitly defined rules of use.

This capability has been designed for use within Virtual Organizations (VO). It is easy to manage in multi-administrative environments, integrates with related VO capabilities (such as the VO-Set-Up capability described in Sect. 1.3.1). For each identity federation context, it represents a partner-specific viewpoint of the associated Circle-of-Trust in a way that trust relationship between identity brokers respect supply relationships associated with the domain.

Finally, it is designed for the in-cloud use—it is equipped with a secure web-services remote management interface that enables it to be assembled and man-aged remotely and provides the basis for an instrumentation layer utilized by collaboration services such as the capabilities described in Sect. 1.3.1.

An overview of the architecture of the Identity Broker is shown in Fig. 1.3. To manage a set of dynamically instantiated services as pluggable modules, the management interface is split into two parts: a set of 'core' management methods and a single 'manage' action that dispatches management requests to dynamically selected modules. The signature of the 'manage' method is parametric and dynamically composed depending on the management interfaces of the modules integrated in a given instance (STS) of this capability. The flexibility of XML and SOA Web Services technology enables this form of dynamic composition.

Referring to Fig. 1.3, the core management methods include operations for creating new federation configurations from given specifications, for temporarily disabling or enabling them and for inspecting their values and meta-data. A proxy function forwards aspect-specific management requests to the management module of the respective provider—i.e. the bundle of process and module implementations fulfilling an aspect of the STS operation in a given context.

Each federation context has an associated federation selector—a mechanism that maps a virtual identity (e.g. security token) issuance request or validation message or a management operation to an STS instance configuration. This can be for example a WS-Trust request for issuing an XML security token, such as a SAML

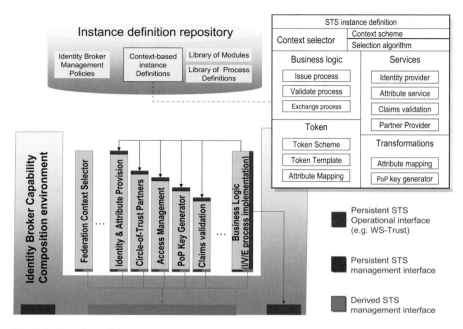

Fig. 1.3 Overview of the Identity Broker architecture

assertion, in the scope of a given collaboration [37]. In a simple case, the federation selector could contain a unique identifier or a collection of WS-federation meta-data [38]. When clients request an STS to issue validate tokens, the STS will determine whether this can be done based on the information it holds in its database. A fault message will be returned to the requestor if no suitable context is identified.

After selecting the matching federation configuration, the identity broker instantiates the corresponding STS business logic capability and binds it with the applicable process description. It also instantiate the internal capabilities of the STS such as the corresponding federation partner provider, the claims provider and the claims validity provider and binds them to the STS business logic process. Each of these internal capabilities of the STS may also have a federation-context-specific configuration, which is loaded upon their instantiation. An innovative execution mechanism by which instance execution takes the form of separate bundles of parallel threads that are allocated distinct memory spaces ensures high-performance during operation.

1.3.2.2 Distributed Access Management

Distributed access control and authorization services allow groups of service-level access policies to be enforced in a multi-administrative environment while ensuring regulatory compliance, accountability and auditing.

Until recently most of the research into access control for networks, services, applications and databases was focused on a single administrative domain and the hierarchical domain structures typical of traditional enterprises. However, the dynamic nature and level of distribution of the business models that are created from an SOI—especially when this incorporates Cloud services—often mean that one cannot rely on a set of known users (or fixed organizational structures) with access to only a set of known systems. Furthermore, access control policies need to take account of the operational context such as transactions and threat levels. The complexity and dynamic and multi-administrative nature of such IT infrastructures necessitate a rethink of traditional models for access control and the development of new models that cater for these characteristics.

The access management capability, developed in the security theme of BEin-GRID, provides a means for specifying policies that control service-level access and usage in such environments and for automating the necessary decision-making while facilitating accountability and security auditing. It can recognize multiple administrative authorities, admit and combine policies issued by these authorities, establish their authenticity and integrity and ensure accountability of policy authoring, including the non-repudiation of policy issuance. The validity of the access policies authored by different administrators is established by means of digital signatures from the policy issuing authority (e.g. the administrator authoring a policy or a recognized authority vetting the administrator) and may be time-limited and must be historically attested.

This access management capability also caters for policies addressing complementary concerns (operational and management) in a multi-administrative environment. It supports policies about:

- Subjects accessing resources in a context, where policies will be issued (and signed) by administrators authorized to manage resources.
- Who can delegate which access rights about which resources in what context.
- Obligations that instruct associated policy enforcement points.

Constrained administrative delegation [28] is a feature that allows some administrative authorities to author (delegation constraint) policies that constrain the applicability of (access) policies authored by other administrative authorities. Constraints may take the form of rules that apply on a subset of the available attribute types and policy evaluation algorithms. This allows, for example, for safely delegating policy management rights empowering customers to manage the rights of their users directly accessing in-cloud resources in the case of multi-tenancy hosting scenarios, common in Data Centers and Cloud Computing.

In all cases, there may not be any prior knowledge of the specific characteristics of subjects, actions, resources and so on. Hence, there are no inherent implicit assumptions about pre-existing organizational structures or resource or attribute assignments. This comes in contrast to access control lists and traditional role-based access control frameworks in several ways:

Attribute schemes and attribute assignment processes may evolve independently of the access policies; different authorities can be in charge of attribute definition, attribute assignment, access policy authoring, and access control.

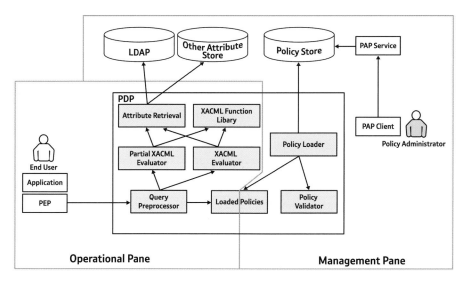

Fig. 1.4 Overview of the architecture of the Distributed Access Management capability

During access policy evaluation, access decisions may consider environmental attributes and other contextual information in addition to attributes of the subject, resource and action. Contextual information evolves during the policy life-cycle.

Policy administration and decision making may also be contextualized. Different administration and/or command structures may manage independent life-cycle models and policy groups associated with different contexts. Access policies may also need to be executed within the scope of a particular context that influences the way in which their evaluation algorithms are being applied.

In some cases, it may also be necessary to ensure segregation of policy execution—that is, that ensure no interference between the policies being executed in different contexts. This capability can create new policy stores and policy engine instances on-demand for use in distinct contexts. This is particularly useful where in-depth process and policy separation needs to be achieved including remote Application Hosting and some Cloud Computing platforms.

The policy decision point (PDP) at the core of the access management capability may be exposed as a hosted service, be deployed as a component of a policy decision making capability with a larger scope (such as a federated identity and access management capability) or be an integral part of the policy enforcement (PEP) function. It is also be possible to deploy the overall access management capability as a managed service, if needed.

In the scope of the BEinGRID project researchers at BT in collaboration with Axiomatics—a SME specializing in policy and entitlement servers that offers access management capability to the Swedish NHS—developed an access management capability that meets the requirements listed above. This development took place in coordination with the development of version 3.0 of the XACML standard at OASIS. This version of the XACML standard is in the process of introducing a number of new features including policy-based constrained administrative delegation and obligation policies. These features have been explored by the BEinGRID capability. An overview of the architecture of this access management capability is shown in Fig. 1.4.

1.3.3 Common Capabilities for Managing Software Licenses

One of the key business issues derived from this elicitation of requirements from the BEinGRID BEs has been the need to improve support for commercial applications from independent software vendors (ISV) in SOI and Cloud computing environments. Small and medium enterprises (SME) from the engineering community especially stand to profit from pay-per-use HPC scenarios akin to Utility Computing.

For example, very few enterprises however maintain their own simulation applications. Instead—in contrast to academic institutions—commercial applications from independent software vendors (ISV) are commonly used with an associated client-server based licensing. The authorization of these client-server based license mechanisms relies on an IP-centric scheme: a client within a specific range of IP-addresses is allowed to access the license server. Due to this IP-centric authorization, arbitrary users of any shared IT resource may access an exposed license server, irrespective of whether or not they are authorized to do so.

In the absence of controlled access to a local or remote license server that is suitable for HPC utility and in-cloud hosting, it is often not possible to use commercial ISV applications in these environments. Consequently, a large number of commercial users have are not able to use ISV applications in such environments.

The LMA (License Management Architecture) capability developed in the License Management theme of BEinGRID is in our knowledge the first complete solution for HPC utility or Cloud platforms that solves this problem. LMA is architected as a bundle of capabilities, shown in Fig. 1.5, which combined enable managing software licenses for shared resource use. One notable innovation has been the ability to transparently reroute the socket-based communication via a SOCKS proxy-chain that is scalable and suitable for supporting legacy and proprietary client-server protocols that are currently used in commercial environments. Another innovation has been a mechanism to authorize access based on one-time credentials (in close analogy to PIN/TAN solutions) that is suitable for use over open infrastructures with varying levels of trust and enable run-time authorization and context-based accounting.

The LMA capability is generic, independent of specific middleware choice, and features a cost-unit based accounting. It enables using licensed ISV applications in HPC utility or Cloud platforms in a wide range of provisioning scenarios. In combination with secure access to the license server, LMA facilitates the non-interruptive

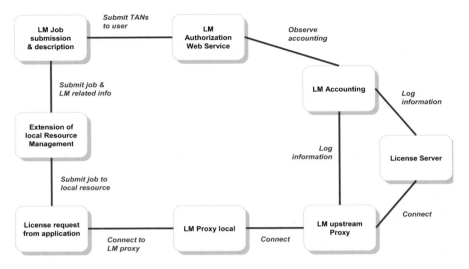

Fig. 1.5 Overview of the architecture of the License Management capability

business transition to pay-per-use models while supporting the current legacy technology that used to manage software licenses. It therefore enables increasing of the market size in the area of SOI and on-demand Cloud computing.

Aspects of LMA have been validated various HPC utility contexts including in a BE that demonstrated the use of a large scale multidisciplinary compute Grid to generate cost-effective and optimized solutions for water management [2], and a BE that demonstrated a solution to reduce the technical and economical risks that are implicit in large and complex ship building projects [5].

1.3.4 Common Capabilities for Managing Service Level Agreements

Quality of Service (QoS) is in essence about a set of quality metrics that have to be achieved during the service provision. These metrics must be measurable and constitute (part of) a description of what a service can offer. The QoS of IT services is often expressed in terms of capacity, latency, bandwidth, number of served requests, number of incidences, etc. The QoS of services offered to the customer is sometimes expressed as a package (for example bronze, silver, gold) and in relation to key performance indicators (KPI). In this case, a match between the elements of the scale and measurable metrics relative to the service is provided.

A Service Level Agreement (SLA) defines the QoS of the services offered. Typically SLA is a formal written agreement made between two parties: the service provider and the service user, defining the delivery of the service itself. The document can be quite complex, and sometimes underpins a formal contract. The contents will vary according to the nature of the service itself, but usually includes a

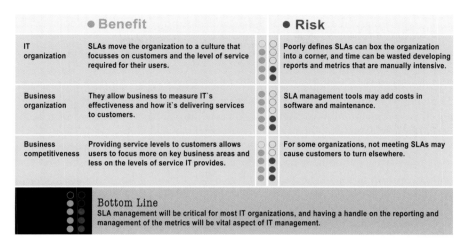

Fig. 1.6 Summary of an impact assessment of SLA use for IT services from [12]

number of core elements, or clauses. These define a specific level of service, support options, incentive awards for service levels exceeded and/or penalty provisions for services not provided, etc. Some organizations, attempting to avoid negative connotations, prefer to use the terms SLE (service-level expectation) or SLG (service-level goal) for the definition of the QoS of the services they offer.

Functional service-level agreements attracted high interest from telecommunication service providers in the late 1990s. As more of the commonly used services are now being delivered online, enterprise, government, and academic environments are moving towards SLA-driven services. However, as was the case with the telecommunications industry in the '90s, the right elements to generate and manage a successful SLA are rarely in place. Many organizations that depend on IT lack the governance structures, service catalogues, defined processes, management and monitoring services that are necessary for managing SLAs successfully.

A successful SLA strategy must include the ability to collect configuration information on network and server assets, access customer information for business impact analysis, and provide data on all internal or external SLAs. Customers should also have visibility of real-time and historic service views. Ensuring that users know what services IT makes available, what level of service is provided, and the ability to verify the level of service offered can help increase customer satisfaction and make the overall relationship between IT and the users radically different from what it is today. To paraphrase a common saying "A service-centric approach to SLA management really means a user-centric approach to the IT offered." Figure 1.6 summarizes an impact assessment of SLA use for IT services.

The BEinGRID project has produced a bundle of capabilities for managing SLAs that enhance common Grid Computing platforms with a comprehensive environment covering the full-life cycle of SLAs for the use of ICT resources and services. One such example includes the first implementation of a comprehensive SLA framework [29] on top of the Globus Toolkit—an Open Source Grid Computing middleware commonly used in large-scale science projects and some commercial applications [22].

Capabilities for SLA *specification* include support for standards-based specification of SLAs and templates, such as [36]. Service delivery is described through the {service, SLA} pair, defining exactly what the client is expecting from the provider. The complete lifecycle of the service is mirrored by the life-cycle of the corresponding SLA specification. As such, the SLA has a lifespan that is at least as long as the period of service usage by the service consumer.

The main challenge of SLA *discovery and negotiation* resides in providing a comprehensive environment for offering and demanding bargaining, in the legal parts as well as in the technical parts. This allows both parties to obtain a contract which is most fit for its use, minimizing over- and under-provision. WS-Agreement [36] is the only standard in this area that has met some acceptance. In March 2009 WS-Agreement has been in its last steps to become a full standard. Several implementations of this specification have been developed since 2007. Some are available as Open Source software from [18]. However the technical means to perform bargaining are not yet there: The WS-AgreementNegotiation protocol is still at early stages of maturity and, although experts in the BEinGRID project have explored some early features of such protocols, their experience of the BEinGRID BEs is indicating that SLA negotiation is applicable merely in relation to SLA discovery. The business reasoning for providing a capability to re-negotiate SLAs remains unclear (as opposed to cancelling an SLA and replacing it with a new one). BEinGRID experimentation confirmed the applicability of either simple short-term SLAs for use of IT resources or of complex legal contracts. The latter are perceived as a means of treating higher value or higher risk offerings by the parties involved, their definition typically involve qualified lawyers and would not be automatically renegotiated. Furthermore national law in some European regions obligates that renegotiation is treated as a negotiation of a new contract. Nevertheless, it appears as if in some cases companies are willing to enter a fixed long-term contract, and allow for short-term contracts (typically referencing the over-arching long-term legal contract) that can be negotiated automatically, within a limited scope.

The SLA *Optimization* capability matches the information offered in SLAs to the available resources. This improves the provider's scheduling strategy, allowing the provider to improve the utilization of its resources. It also allows implementation of the business rules which govern the allocation of resources based on KPIs such as the return value of the incoming SLA requests. Most schedulers are designed to optimize the resource usage based on the incoming resource requests, but very few take into account KPIs such as the business value of the request.

The SLA *Evaluation* capability compares information collected from sensors and other monitoring tools to the SLA objectives, and raises alarms when thresholds are passed or constraints are violated. The provider, having detailed information of its resource status, can act proactively to address failures, thus managing the risk associated with the penalties incurred. The consumer may also receive such notifications, and can reallocate tasks, enhancing its ability to react to the likelihood of failures. This capability builds on a modular architecture that exploits a topics-

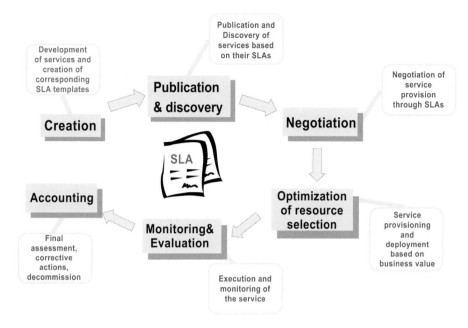

Fig. 1.7 Common capabilities for SLA management against the life-cycle of managing SLAs

oriented publish-subscribe model and can instrument native or "off-the-self" ICT re-source monitoring tools. Intelligent event correlation and non-repudiation combined with SLA-based accounting and annotation of violations make information clearer and reliable enough for enabling evidence gathering and evidence-based decision making regarding claims for compensation. Early developments of this capability stemmed from collaborative research in the TrustCoM project [15].

The *SLA-based accounting* capability supports the selection and adoption of the suitable charging scheme for each service execution environment based on the metrics included in the SLA specification. Such metrics may deal with a variety of heterogeneous resources. This capability enables charging for service use based on its real execution cost. The analysis reports produced by this capability also helps clarifying resource usage and causality of retribution and penalties.

Figure 1.7 shows an overview of the capabilities developed by the BEinGRID project against the typical life-cycle of managing SLAs.

The business benefits of such a comprehensive environment for SLA manage-ment over Grid middleware include optimizing resource allocation and use in re-sponse to market requirements, reducing Total Cost of Ownership (TCO) by im-proving efficiency of resource utilization and faster and better targeted response to failures, increasing customer confidence by allowing transparency of operation (subject to the SLA), and enabling customizable billing by providing finer granu-larity of accounting and reporting. These results have been validated in Business

Experiments in different vertical market sectors including the pilot projects in the area of online collaborative gaming [3] and eHealth [9] mentioned earlier.

1.3.5 Common Capabilities for Data Management

The analysis of main technical business issues faced by the European industry by the BEinGRID projects identified common concerns about storage, access, translation and integration of data in most market sectors. These can be simplified in the following key points—further analyses are included in Chap. 7 of this book.

- Where should I put my data?
- How should I get to it?
- How do I present my data in a way others will understand?
- How can I combine data from different places?

All of these questions are important to modern businesses. In many industries, collaboration and the efficient flow of information between organizations is critical. For example, Just-In-Time techniques [33] aim to improve the efficiency of a supply chain and to do this effectively they need access to up to date information from multiple organizations.

The focus of the Data Management theme of BEinGRID has been to address the challenges of accessing, integrating and utilizing existing data that may be heterogeneous and originate from multiple business partners in a value network.

There main results from this thematic area include capabilities for facilitating access to remote data sources, for homogenizing the treatment of data sources, and for synchronizing multiple data sources. Reference implementations of these have been developed over OGSA-DAI platform. The latter is contributed by the Open Grid Services Architecture—Data Access and Integration (OGSA-DAI) project [25], a part of the Open Middleware Infrastructure Institute UK [26].

More specifically the following common capabilities have been identified and developed over OGSA-DAI:

- Data Source Publisher: This capability simplifies the set-up of existing grid middleware by allowing a source of data to be published over web services. It also reduces the cost of adopting OGSA-DAI.
- OGSA-DAI Trigger: This capability enhances OGSA-DAI with new data integration features and allows for automated data integration using OGSA-DAI. Underpinning this capability is innovation that allows executing an event-driven OGSA-DAI workflow when a database changes.
- JDBC Driver: This capability offers a new interface for OGSA-DAI that allows enhanced data integration in existing applications and makes integrated data resources appear as a simple database.
- OGSA-DAI SQL views: This capability allows adapting an existing data source for use in a Data-Grid; it enables a view that is independent of the data source and appropriate for use in a Data-Grid without affecting the original data-source.

The results in the Data Management area offer new opportunities for collaboration between business partners by enabling access to sources of information, reducing costs due to better integration of data across sites and enabling the development of simpler data-oriented applications. They also improve the OGSA-DAI framework with a more comprehensive data integration capability and reduce the barriers to adopting OGSA-DAI in business environments.

The results in the Data Management have been validated in a BE demonstrating the use of Data-Grid technologies for affordable data synchronization and SME integration within B2B networks [8]. Some aspects were also validated in a BE demonstrating improvements to the competitiveness of textile industry gained by implementing a SOI between textile firms and technology provider that focused on offering high end services such as production scheduling, global resource scheduling and virtual retailing. Other aspects were validated in a BE focusing on supporting post-production workflow enactment in the Film industry [1].

1.3.6 Common Capabilities for Data and Service Portals

Portals are commonly used as a means of obtaining a unifying view of SOI and Cloud platforms and of introducing transparencies that hide the complexity of the underlying IT infrastructure. They include portals for managing user communities, portals for accessing distributed data sources and portals for managing the life-cycle of computational tasks (i.e. submitting, monitoring in real-time and controlling a job). Many of the 25 BEinGRID BE had business needs relating to the use of such portals. Based on the analysis of their requirements, the strongest business needs for technological innovation in this area were organized in three sub-categories:

- Security, user provisioning and user management.
- Efficiency and security of file and data sharing.
- Visibility and manageability of submitting, monitoring and controlling transactions, jobs and other computational tasks.

Typically, such business needs become even more critical in the case of cross-organizational portals—i.e. portals shared among a community of business partners (Virtual Organization), portals that offer access to shared resources, or portals that offer access to federated services or resources offered by a Virtual Organization. Unfortunately, this is where most current solutions appear to be weaker.

The main research and development results in this area have taken the form of extensions to a Plug & Play portals development framework built on top of the Open Source Vine toolkit [35] as presented in Chap. 8 of this book. The key innovations underpinning this result are a configurable abstraction layer that uses Web2.0 mash-up technology to hide complexity of Grid Computing tasks, and an innovative user and account provisioning mechanism.

This framework helps in reducing integration costs and preserve existing investment by facilitating integration with existing solutions through a flexible plug-in

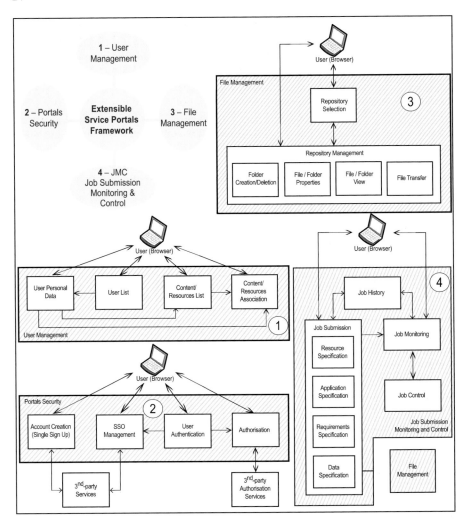

Fig. 1.8 Aspects of an extensible service portals framework for Grid and Cloud computing

adaptor mechanism. Ease of integration with existing content management tools and legacy applications also results in reducing the cycle time of Grid portal development projects. Finally the user provisioning and administration mechanisms help reduce human error, coordinate application-specific accounts and authentication mechanisms and results in an easier to manage uniform administration layer. The high-level architecture of the main capabilities developed for this framework is shown in Fig. 1.8.

Further studies [13] and [27] have analyzed how this framework can be further enhanced through its integration with other capabilities developed by BEinGRID.

Report [13] explores the added value of integrating this framework with the federated identity and access management capabilities mentioned in previously this chapter. Report [27] explores the added value of integration with the License Management capabilities mentioned in previous sections of this chapter.

Results in the area of Grid Portals have been validated in various vertical market sectors including a Business Experiment focusing on production scheduling and virtual retailing in the Textile Industry [4] and a Business Experiment demonstrating the enactment of Web2.0 workflows for Service Oriented Infrastructures in complex enterprises [7].

1.4 An Example that Brings it all Together

The European IT Infrastructure Management Services market was worth almost 50 billion Euros in 2006 according to a report from IDC [23] and has been increasing by almost 10% a year until 2009. It appears that a similar trend is now emerging in the Cloud computing area. Merrill Lynch [24] derives the spending on Cloud computing from total software spending. For 2011, it is expected that 20% of spending on enterprise applications and infrastructure software and 8% of spending on custom software will be spent on Cloud computing. The worldwide Cloud computing market is expected to reach $95 billion by 2011. This represents 12% of the total worldwide software market.

One of the recurrent challenges for businesses in this area is how to manage the deployment, distribution and configuration of the capabilities and resources required for offering a service that is distributed over multiple hosts that may not be under the control of the same enterprise. According to the analysis of the BEinGRID BE, the top four concerns in this area have to do with how to define and enforce security policy, how to measure and optimize resource usage, how to monitor and evaluate the quality-of-service offered against a Service Level Agreement (SLA), and how to manage configuration over a federation of hosting platforms.

In response to this challenge, and as an illustration of combining in practice many of the common capabilities developed, the BEinGRID project has developed a common capability called "*(Enhanced) Application Virtualization*". This capability enables the management of the deployment, distribution, coordination and configuration of capabilities and resources required for offering applications distributed over a federated set of network hosts. It can be used to add an instrumentation layer and coordinate different service execution environments in order to enable the secure and manageable application exposure of remotely hosted (and potentially distributed) applications. A differentiator compared to what is currently available in the market is that this collection of capabilities provides a unifying layer for managing security (i.e. identity, access, secure service integration), SLA fulfillment and performance monitoring across multiple platforms.

An evolution of this bundle of capabilities could also be exploited to coordinate the integration of, and manage, software-as-a-service (SaaS) offered on Cloud platforms of different providers (e.g. Amazon, Microsoft Azure, etc.). It is reasonable, in

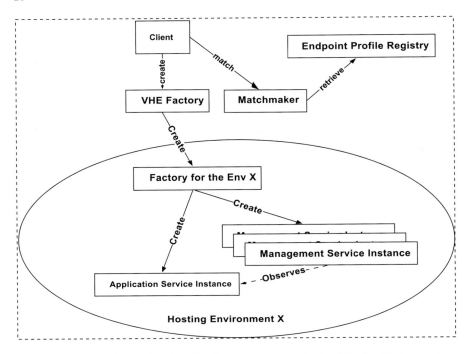

Fig. 1.9 Creation of in-cloud SaaS application instances on an in-cloud Hosting Environment

fact, to suppose that different Cloud providers could differentiate their offers hence generating a market where different Cloud platforms are best fit for hosting different kinds of services. Consequently offering a capability enabling the selection of most suitable providers for hosting a SaaS solution as well as coordinating application deployment and exposure on Cloud platforms offered by different providers can be attractive and produce high return on investment. According to a 2009 survey of European SMEs by ENISA [16] the majority of responders (32%) consider a federation of Cloud platforms offered by various providers to be most suitable Cloud for an SME. A close second (28%) is a Cloud platform offered by a trusted partner for use by a business community.

A typical usage scenario of this capability is shown in Fig. 1.9, where an Application Service Provider (ASP) provides an in-cloud application to a client on the basis of an agreed contract (SLA). In order to optimize capital expenditure and to match use of IT resources to business demand the ASP has joined a community of Cloud platform providers that can offer the resources, platform and infrastructure services that the ASP needs in order to provide this in-cloud application as SaaS to its own user community. In order to monitor service usage and optimize resource utilization the ASP creates an instance of the application for each customer it serves based on Quality of Service parameters that reflect the corresponding customer agreement (SLA). A separate reference to a service endpoint is produced for each instance of the application. The creation of the application instance—i.e. the "application

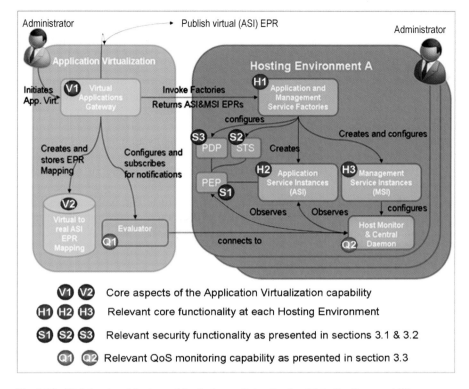

Fig. 1.10 High level architecture of the (enhanced) Application Virtualization capability

virtualization"—is initiated via the ASP via Cloud service management interfaces that are offered by the Cloud platform federation (represented by a Broker).

The ASP is assured by the Broker (representing the Cloud platform federation) that the created instance can meet the SLA it has agreed with its customer and is provided with the necessary capabilities for managing the life-cycle of the application instance and the policies governing the (virtual) service delivery platform through which the application is offered to the ASP's customers. The ASP is not exposed to the complexity and heterogeneity of the capabilities that have been combined in order to allow the application service delivery. Unless described in the SLA, the ASP avoids exposure to the specifics of where specific application resources have been deployed. The ASP has delegated to the community of Cloud platform providers (represented by a Broker) the selection of suitable hosting environments and the execution of processes that implement the deployment and configuration of application instances and their exposure as a service. It has been provided with specialized management services that the ASP uses for coordinating these processes and managing the operation of the in-cloud application services it offers to its customers throughout their life-time.

The virtualized application is exposed via an in-cloud service Gateway and the integration of any other value adding services (VAS)—potentially provided by third parties—catering for the non-functional aspects of the application is transparent to the application consumer. The capability enables the ASP to use standardized management services in order to govern the configuration of the virtualized application, the underlying virtual service delivery platform and any third party value adding services (VAS) such as SLA and security capabilities that have been selected by the ASP to enrich the customer experience. The adoption of the Gateway offers the necessary location and platform transparency while acting as an integration point (i.e. a virtual service bus) to external value adding services.

In terms of business impact, this bundle of capabilities allows an ASP to offer their applications as a service in a simple and manageable way without being exposed to the detail of managing the enabling infrastructure. This increases flexibility and allows a separation of concerns between application provisioning and management, and facilitates their transition towards a SaaS model.

1.5 About the Contents and Structure of this Book

This book targets a general audience of strategists, technical consultants, researchers and practitioners in Service Oriented Infrastructure technologies with emphasis on Grid and Cloud Computing.

In this chapter we presented some business challenges that are common among several sectors of the European market and then we summarized a selection of common capabilities (i.e. services capturing reusable functionality of IT solutions) for Service Oriented Infrastructures and Cloud Computing that can be used to address these challenges. These capabilities have been developed as part of the technological research and innovation activities of BEinGRID and they embody technological innovations in several areas that are considered to be critical for the way that business will be conducted in the future, based on experience with 25 pilot projects that cover many sectors of the European economy. These pilots also offered real-life business scenarios and a platform for validating the common capabilities and for identifying best practices in close liaison with stakeholders in value chains that are representative of each vertical market sector and the European economy as a whole.

In our analysis, we highlighted the likely *impact of innovation* produced by each common capability, and referred to concrete examples of publicly available descriptions of pilot projects and real-life business scenarios where the current state-of-the-art can be improved by exploiting implementations of these common capabilities. In each case, our analysis included a reflection of the interaction between the technical experts innovating, the business analysts supporting them and a relevant pool of business stakeholders. Such analysis and validation of technological innovation is of an unprecedented size and diversity not only in the history of European research and innovation but also globally.

In this chapter we also presented a futuristic, indicative *integration* scenario that illustrates how several reusable capabilities that originate in diverse thematic areas, and meet diverse business requirements can be brought together in order to solve a challenging and complex problem that may appear as the market evolves.

This chapter only covers at a high level a small subset of the common capability definitions, the associated design and implementation patterns and the validation scenarios that constitute the main body of knowledge and recommendations produced by the technology innovation stream of BEinGRID. The remaining chapters of this book provide a more extensive analysis of these results.

In Chaps. 3–8 we present the selected innovations in each thematic emphasizing examples of their use and anticipated business impact. More specifically:

Chapter 3 focuses on results that enable VO Management. These results include technical innovations that help businesses to establish secure, accountable and efficient collaborations and operate services and resources securely over a shared ICT infrastructure. This is embodied in capabilities that enable the life-cycle management of VOs, the discovery of resources within a configured VO, and the configuration and operational management of virtual service platforms that can host distributed applications offered as a service by a VO.

Chapter 4 focuses on results that enable securing business operations in an enterprise and across partners in a VO. These results include technical innovations for securing the services exposed by an enterprise, for managing the establishment of trust and federating identities among business partners, for managing access to services and resources in multi-administrative environments, for monitoring security events, and for governing the life-cycle of security policies (for identity, access, service-service communication, and monitoring) and their enforcement on a distributed service oriented infrastructure. These results are embodied in a collection of capabilities that enable securing the operation of VOs as well as securing multi-tenancy hosting of services in a distributed infrastructure. They underpin or compliment solutions offered in VO Management and other thematic areas.

Chapter 5 focuses on results that introduce new functionality or improve aspects of a License Management Architecture that sufficient for enabling "pay-as-you-go" (PAYG) license models over a Grid executing jobs submitted are the network by a customer or a Cloud Computing platform hosting a customer's services. These innovations enable the adoption of business models akin to PAYG that are suitable for utility computing and have been identified as a significant business opportunity across several vertical markets.

Chapter 6 focuses on results that improve several aspects of the SLA management life-cycle on a distributed SOI. These include improvements to open standard schemes for specifying SLAs, capabilities for selecting the most appropriate hosts in a distributed infrastructure in order to optimize resource utilization, and capabilities for accounting and fine-grained monitoring of performance and resource utilization against an SLA.

Chapter 7 focuses on results for managing storage, access, translation and integration of data on a distributed SOI. These include technical innovations for aggregating heterogeneous data sources in virtual data-stores, for optimizing and ensuring seamless access to distributed and heterogeneous data sources.

Chapter 8 focuses on results for service and data portals that enable scalable solutions providing an instrumentation layer with intuitive interfaces for managing user communities and complex processes and data in SOI. These include the innovative use of Web2.0 technologies for hiding the complexity of SOI and improving customer experience by creating a simper working environment for the end user and the ICT infrastructure managers alike.

Chapter 9 presents examples of hypothetic scenarios where a large selected innovations explained in previous chapters have been used together in order to address a challenge that has been recognized as becoming increasingly important for businesses with the uptake of Cloud computing.

More information about the innovations in each area including white papers, training material, demonstration videos and software is provided at [21].

References

1. BE02—Business Experiment 2: Movie post-production workflow, http://www.beingrid. eu/be2.html
2. BE06—Business Experiment 6: Ground water modelling, http://www.beingrid.eu/be6.html
3. BE09—Business Experiment 9: Distributed online gaming, http://www.beingrid.eu/be9.html
4. BE13—BEinGRID Business Experiment 13: Virtual laboratory for textile, http://www. beingrid.eu/be13.html
5. BE16—Business Experiment 16: Collaborative ship building, http://www.beingrid.eu/ be16.html
6. BE22—Business Experiment 22: Grid technology in the agricultural sector, http://www. beingrid.eu/be22agrogrid.html
7. BE23—Business Experiment 23: Workflows on Web2.0 for Grid enabled infrastructures in complex Enterprises (WOW2GREEN), http://www.beingrid.eu/wow2green.html
8. BE24—Business Experiment 24: Grid technologies for affordable data synchronization and SME integration within B2B networks—GRID2(B2B), http://www.beingrid.eu/be24.html
9. BE25—Business Experiment 25 (BE25), Business Experiment in Enhanced IMRT planning using Grid services on-demand with SLAs (BEinEIMRT), http://www.beingrid.eu/be25.html
10. BEinGRID Booklet, BEinGRID—Better Business Using Grid Solutions—Eighteen Successful Case Studies Using Grid, booklet available at http://www.beingrid.eu/casestudies.html
11. BEinGRID Business Experiments in Grid, http://www.beingrid.eu/. Accessed 1 July 2009
12. M. Biddick, Service-level agreements come of age, in Information Week issue 1st December 2008. Also available online at http://www.informationweek.com/
13. D. Brossard, S. Karanastasis, A note on integrating common capabilities from Security and Portals July 2009. To appear at IT-tude.com as a Technical White Paper http://www. it-tude.com/technicalwhitepapers.html
14. T. Dimitrakos, Towards a trust and contract management framework for dynamic virtual organisations, in eAdoption and the Knowledge Economy (IOS Press, Amsterdam, 2004)
15. T. Dimitrakos, M. Wilson, S. Ristol, TrustCoM—A Trust and Contract Management Framework enabling Secure Collaborations in Dynamic Virtual Organisations, ERCIM News No. 59, October 2004. See also the TrustCoM project web site: http://www.eu-trustcom.com
16. ENISA (2009) An SME perspective on Cloud Computing—Questionnaire, June 2009. For more information please contact ENISA via http://www.enisa.europa.eu/
17. Gartner (2008) Assessing the Security Risks of Cloud Computing, 3 June 2008
18. GRAAP—Web site of the Grid Resource Allocation Agreement Protocol Working Group, Open Grid Forum (OGF), http://forge.gridforum.org/sf/projects/graap-wg
19. IT-tude.com Technology Solutions (2009) VO Management, http://www.it-tude.com/vo-article.html. Accessed 1 July 2009

20. IT-tude.com Technology Solutions (2009) Security, http://www.it-tude.com/grid-security.html. Accessed 1 July 2009
21. IT-tude.com (2009) From Service-oriented IT to Business Value, http://www.it-tude.com. Accessed 1 July 2009
22. GT4—Globus Toolkit version 4, http://www.globus.org/toolkit/
23. IDC (2005) European Infrastructure Management Services Market, Forecast and Analysis, 2005–2009, IDC April 2005
24. Merrill Lynch (2008) The Cloud Wars: $100+ billion at stake, Research note, May 2008
25. OGSA-DAI (2009) Web pages at http://www.ogsadai.org.uk. Visited 14 April 2009
26. OMII-UK (2009) Web pages at http://www.omii.ac.uk/. Visited 14 April 2009
27. Y. Raekow, C. Simmendinger, P. Grabowski (2009) An Improved License Management for Grid and High Performance, July 2009. To appear at IT-tude.com Technological Solution White Papers http://www.it-tude.com/technicalwhitepapers.html
28. E. Rissanen, B.S. Firozabadi, Administrative delegation in XACML, in Proceedings of the W3C Workshop on Constraints and Capabilities for Web Services, USA (2004)
29. I. Rosenberg, A. Juan, The BEinGRID SLA framework. Report available at http://www.gridipedia.eu/slawhitepaper.html (2009)
30. SAML, Assertions and Protocols for the OASIS Security Assertion Markup Language (SAML) V2.0. OASIS Standard, March 2005, Document ID saml-core-2.0-os
31. J. Staten et al., Which Cloud Computing Platform Is Right For You? Understanding The Difference Between Public, Hosted, And Internal Clouds, Forrester, 13 April 2009
32. The Economist (2008) The long nimbus, The Economist, 23 October 2008
33. Toyota MMK—Toyota Production System Terms (2009) Toyota Motor Manufacturing Kentucky, http://toyotageorgetown.com/terms.asp. Visited 14 April 2009
34. UDDI, OASIS UDDI v3 Specification TC, http://uddi.org/pubs/uddi-v3.0.2-20041019.htm
35. Vine—IT-tude.com Technological Solution: Vine Toolkit, http://www.it-tude.com/thevinetoolkit.html
36. WS-Agreement—Web Services Agreement Specification, Grid Resource Allocation Agreement Protocol (GRAAP) WG, Open Grid Forum, 14 March 2007
37. WS-Trust, OASIS WS-Trust specification v1.3, http://docs.oasis-open.org/ws-sx/ws-trust/200512
38. WS-Fed, Web services federation language v1.1, IBM December 2006

Chapter 2
Approach Towards Technical Innovations for Service Oriented Infrastructures and Clouds

Theo Dimitrakos, Angelo Gaeta, and Craig Thomson

Abstract The chapter describes the approach used by BEinGRID for delivering technical innovations for Service Oriented Infrastructures. In Sect. 2.2 the methodology to produce the technical innovation is introduced, then the fundamental concepts, Technical requirement. Common Technical Requirements, Common Capability, Generic Components and Validation Scenario are defined and discussed. In Sect. 2.3 the knowledge flow in the project between the different technical and business expert teams to achieve an innovation delivery process is presented. The elicitation of the common technical requirements and the thematic are deeply introduced in the following sections. The modelling phase with prioritisation of common technical requirements, common capabilities, design templates, design patterns and generic components are parts of the workflow and described in Sects. 2.3.3 and 2.3.4. At the end of the chapter an introduction to the structure of the book is given.

2.1 Introduction

In this section we present the methodology that has been used in order to produce the technical innovations summarised in the remaining of the book. The results are the outcome of an effort of small teams of three to five researchers and technology experts per thematic area who focused on the definition, design and demonstration of services and components that solve significant common challenges faced by businesses that are adopting SOI technologies. The technology designs and implementations of these results are commercially exploitable and—to a large extent—platform neutral. For a period of three years these teams have been overseeing, analysing and consulting a large community of researchers and practitioners focusing on innovative applications of SOI technologies in various vertical market sectors, and interacted with business analysts who conducted complementary market, legal and economical studies of the same community.

T. Dimitrakos (✉)
Centre for Information and Security Systems Research, BT Innovate & Design, PP13D, Ground Floor, Orion Building, Adastral Park, Martlesham Heath IP5 3RE, UK
e-mail: theo.dimitrakos@bt.com

T. Dimitrakos et al. (eds.), *Service Oriented Infrastructures and Cloud Service Platforms for the Enterprise*,
DOI 10.1007/978-3-642-04086-3_2, © Springer-Verlag Berlin Heidelberg 2010

Table 2.1 The fundamental concepts

Concept	Definition
Technical requirement	A technical requirement is a singular documented need of what a particular product or service should be or do in the specific context of a pilot project. It is a statement that helps to identify a necessary attribute, capability, characteristic, or quality of a system in order for it to have value and utility to a user.
	Requirements show what elements and functions are necessary for a pilot project in a particular vertical market sector.
Common technical requirements	A common abstraction of a collection of related requirements stemming from pilots in several vertical-market sectors.
	These common requirements capture the essence of several requirements. They abstract away the specific application context and seek to meet business or technical challenges that underpin the applications of SOI technologies in several vertical market sectors.
Common capability	A named piece of functionality (or feature) that is declared as supported or requested by an agent. It can be shown to address one or more common technical requirements and to provide a building block for products or services that address related technical requirements in different business contexts.
	A similar concept of a common capability was proposed in [11] to address a major challenge for any information technology and telecommunications company is how to improve time to market for new services and at the same time reduce costs and improve the customer experience. This problem has been tackled at a very large scale with initiatives such as BT's 21 Century Network. The result of the analysis was to propose the concept of reusable capabilities—a concept widely used in the manufacturing industry and now applied to complex telecommunications services to meet these challenging goals.
Generic component	The reference implementation of a common capability as a network-hosted service or as a reusable middleware component on a selected platform.
Validation scenario	Validation confirms that the needs of an external customer or user of a product, service, or system are met. It is the process of establishing evidence that provides a high degree of assurance that a product, service, or system accomplishes its intended requirements. This often involves acceptance of fitness for purpose with end users and other product stakeholders.
	Validation scenario are use-cases and demonstration scenarios agreed with one or more business experiments that demonstrate that the needs of the customer of the implementation of a common capability are met in the specific business context of the pilot project and the corresponding vertical market sector.

2.2 High-Level Objectives

The fundamental concepts (Table 2.1) are used in the remaining of the chapter.

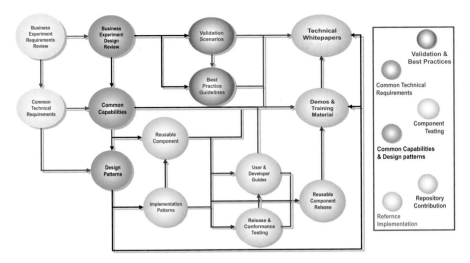

Fig. 2.1 Research and innovation outputs associated with each technical innovation

The main objectives of the BEinGRID approach towards technical innovations have been to produce technical and business innovations that are required in order to maximise the commercial potential of service oriented infrastructures, grid and cloud computing across vertical markets.

These innovations take the form of core, generic functionality or processes that can be implemented over various commercial and experimental service oriented middleware and infrastructures in order to add or help realise business value that is evidenced to be important for commercial success. They build on the experience and work done by the expert teams embedded in the pilot projects across vertical market sectors (BEinGRID Business Experiments) and used some of these pilot projects in order to demonstrate and validate the technology innovation in a business context.

These innovations have been delivered by the following outputs (summarised in Fig. 2.1):

1. Producing *common technical requirements* that identify specific challenges where technical innovation is required. These requirements have been *elicited* by analysing several BE across vertical market sectors;

   – Their *interdependences* have been analysed within and across thematic areas; and
   – They have been prioritised in terms of *innovation potential* and anticipated *business impact* based on feedback from BE in several market sectors and *criticality*[1] in terms of the identified interdependences.

---

[1] In simple terms, criticality of a technical requirement is a function of the number and relative priority of other requirements that depend upon it.

2. Describing *common capabilities* that capture the generic functionality that would need to be in place in order to address these requirements.
3. Producing *design patterns* that describe one or more possible solutions that describe how systems may be architected in order to realise each common capability.
4. Producing *reference implementations* that realise selected common capabilities over commercial middleware. These have passed a quality assurance process that includes:

   - *Release testing* focusing on robustness, installation and usability of artifacts.
   - *Conformance testing* to assure that the artefacts are adequately implementing the functionality of the corresponding common capability.
   - *Documentation* and *training material* explaining how to deploy, integrate and improve the artifacts.

5. Producing *integration scenarios* illustrating how a critical mass of interdependent common capabilities can be implemented together to maximise added value.
6. Producing *validation scenarios* illustrating the benefits of implementing selected common capabilities to enhance business solutions in real-life case-studies.
7. Producing *best-practice guidelines* explaining how these common capabilities can be taken advantage of in indicative business contexts.
8. Various *auxiliary content* such as technical reports, white papers, presentations, demonstration videos and training material that is made available through the Technological Solution and Business Value sections of the IT-tude.com website [6] (also known as Gridipedia).

2.3 The Innovation Delivery Process

The following Fig. 2.2 describes the main dependencies and knowledge flow between the teams of technical experts and business analysts and the BE.

The starting point for research and innovation has been the prior expertise of common research problems in several areas of Service Oriented Infrastructures in general and Grid and Cloud Computing in particular. This includes trust and security, service management, data management, virtual organisation, collaboration management and portals. The technical experts are familiar with SOA Web Services and Grid and Cloud Computing technology platforms, with design patterns and best practices and with innovative solutions that stem out of academic and commercial research in the area.

The starting point for the BE is complementary: the team in each BE brings prior knowledge of the business drivers and technical problems in their specific application domain and vertical market. They include end-users and providers who are familiar with the specific business models and technology innovation drivers for the domain of the BE. These teams also include integrators and technology providers who have expertise about the specific business applications and technical challenges faced in the specific area of the BE. They are broadly aware of the core technology

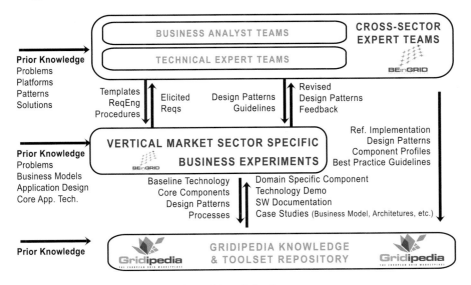

Fig. 2.2 Overview of the dependencies and knowledge flow

but may be sceptical or not informed about new enabling technology which can be tailored and exploited to facilitate optimal solutions in their domain.

The BEinGRID programme leverages the expertise of the teams of technical experts and business analysts in order to define templates and guidance for identifying common problems and eliciting requirements based on software and systems engineering principles, for inspecting architectural designs and eliciting common functionality (common capabilities), procedures and methodological analysis. It also steers the production of vertical market requirements and the solution designs by the teams in each BE. The technical expert teams then analysed these in order to elicit common requirements. The programme also leverages the expertise in the technical expert teams in order to define templates and guidelines for eliciting design information from each BE. These provided a foundation for the architecture of the solution demonstrated by each BE in the corresponding vertical market sector.

In addition to supporting the successful pilots in the vertical market sectors, the interaction with the business experiments provides enough information and insight to the technical expert teams for identifying common technical challenges and producing design patterns that elaborate solutions to these common technical challenges. This regular interaction also contributes to the development of reference implementations of common capabilities for selected vertical market pilots as well as best-practice guidelines about how the design patterns can be applied in the respective business application context. These design patterns, common capability implementations, and best-practice guidelines and case studies have been made available to the vertical market pilots for validation and they are being published to the wider research community via IT-tude.com [6] (also known as the Gridipedia knowledge repository).

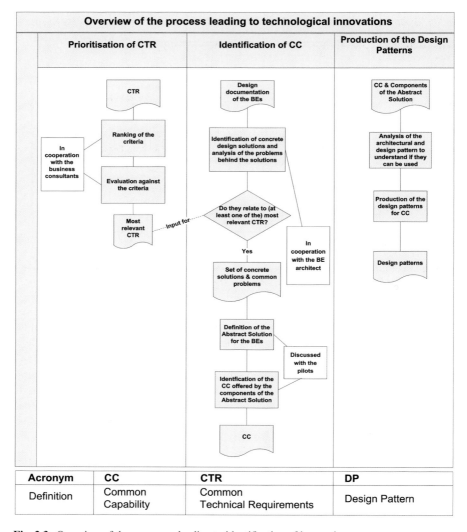

Fig. 2.3 Overview of the processes leading to identification of innovations

The process followed towards producing technical innovations is described in Fig. 2.3. It can be divided in four major stages and two iterations—the first iteration took place in the period June 2006–May 2008 and the second iteration took place in the period June 2008–May 2009:

- Elicitation of common technical requirements.
- Identification of common capabilities and associated design patterns.
- Production of generic components.

• Validation of common capabilities and production of best practice guidelines.

We summarise each of these stages in the following subsections.

2.3.1 Elicitation of Common Technical Requirements

The elicitation of common requirements has been a major task that brought together the technical expert teams with the teams working selected BE. The analysis of requirements took place in four stages.

Preliminary Analysis of Requirements per Vertical Market Pilot

At first, the teams of technical experts produced a requirements template that was filled by each BE. Then the corresponding teams of technical experts analysed the completed templates engaging in a discussion with representatives of the relevant pilots. This analysis was necessary in order to better understand the key technical requirements in each vertical market sector and their relevance to the thematic area of interest for each expert team. This analysis resulted in a classification of technical requirements in each thematic area and an analysis of their dependences. This was complemented by an analysis of market forces and business models in each vertical market sectors that is conducted by the business analyst teams working together with the technical experts.

Elicitation of Common Technical Requirements

Following this preliminary analysis and classification, each team of technical experts focused on further analysing those requirements of the BE classified in their thematic area. During this analysis, the common challenges across BE were identified and common technical requirements were elicited. These common requirements capture the essence of several requirements mentioned by several BE in different vertical market sectors, they abstract away the specific application context and seek to meet generic challenges that underpin common technical challenges.

Analysis of Dependences

Then the dependencies between these common technical requirements were identified and analysed. A further dependency analysis of common requirements across thematic areas has been conducted.

The importance of eliciting common requirements and core technical challenges, from so many business pilots in different market sectors is paramount for this programme of research and innovation Notwithstanding the importance of such results, the added value of a common requirement for a BE is limited unless it is clear how this common requirement and the associated solution (common capability) apply in the context of the specific BE. In order to ensure that common requirements are traceable, the technical expert teams classified the common requirements with respect to the business drivers, the technical challenges and the context of the BE that they relate to. Such a classification and traceability were possible due to the process put in place for eliciting common requirements.

Generic Use Cases

Following the elicitation of common technical requirements, the team of technical experts in each thematic area contributed a collection of generic use case scenarios that illustrates the use, relevance and interdependencies between the common requirements and indicates the common capabilities that may be needed in order to solve the technical challenges that have been identified. These generic use case scenarios have driven the elaboration of common capabilities and the production of design patterns and subsequently generic services and components that implement these common capabilities. Similarly to the requirements elicitation particular emphasis has been put in ensuring that from aspects of the generic use cases one can trace relevant aspects in the specific application scenarios of the corresponding vertical market pilots.

Selected common technical requirements are summarised in Chaps. 3–8 of this book. A more extensive list of these requirements is available at the Technological Solutions part of IT-tude.com [6] (also known as the Gridipedia knowledge repository).

2.3.2 Thematic Areas

The elicited requirements are grouped in the following interdependent thematic areas that are layered as described in Fig. 2.4: higher layers focus more on processes and knowledge representation, while middle layers focus more in application services and information presentation/exchange and lower layers focus more on data, resources and communication.[2]

The *VO Management* thematic area focuses on common administration and governance issues in Service-Oriented Infrastructures that span across organisations. These include:

- *VO creation and management*, i.e. models, processes and technologies for the creation and management of Virtual Organisations.
- *Security in VO*, i.e. federation of trust domains, identity management and cross-organisational access to information and resources.

---

[2]Of course a layering, such as the above offers a simplifying view that is further refined subsequently in this chapter and at the technical part of the Gridipedia knowledge repository by presenting and classifying concrete requirements and analysing their interdependencies.

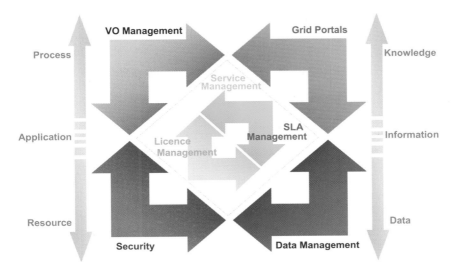

Fig. 2.4 Overview of layered thematic areas and their high-level dependencies

- *Semantics in VO*, i.e. models, processes and technologies for annotation and automatic discovery of shared resource and information within the VO.

The *Security* thematic area focuses on tackling core distributed systems and information or communication security problems that underpin Service Oriented Infrastructures, in general, and Grid computing environments in particular. This area underpins aspects of VO Management relating to federation, VO membership management and federated identity management, as well as security aspects of Portals such as account management (e.g. "single sing-up") and federation of authentication (e.g. "single-sing-on"). It also complements aspects of SLA Management relating to the authenticity of the agreement, the collection, retention and distribution of evidence about performance, and service usage and access control. It also complements aspects of License Management that relate to license authentication, temporary token issuance and validation, and managing access to resources. Finally it complements data management by looking into issues relating to data and information integrity, data confidentiality, secure data retention and transfer. Issues relating to message-level communication security, federated identity and authentication, distributed access control, data and information protection lie at the core of this area.

The *Service Management* thematic area has a fairly broad scope ranging from manageability aspects (e.g. as in applications of the Web Services Distributed Management framework—WSDM) to SLA management and from service provisioning models to License Management. In order to focus efforts, and following consultation with the BEs, we concluded that the sub-themes of *License Management* and *SLA Management* were of highest priority based on the likely innovation dividend and business impact of the anticipated results.

- *License Management* is relevant to several vertical market sectors from the Gaming Industry to highly specialised numerical simulation codes from Independent Software Vendors (ISV). The current support for effective License Management in Grid and Cloud computing environments is particularly poor. License Management was lacked adequate attention in Grid computing partly because of the academic origins of Grid middleware where applications either did not require tight control on licenses or where application licenses were used in closed environments subject to overarching bulk licensing agreements. In contrast, corporate environments typically rely on commercial code basis of ISVs with an associated License Management capability—usually FlexNet from Macrovision, which is the quasi standard in this area. However, business models of most ISV's in this area are based upon customers paying for a fixed number of licenses over a fixed time interval. On average those licenses are only used during a small part of this interval. The logical next step in License Management therefore would be a pay per use model, which implies a big change in the business models of most ISV's, and one that many ISVs indicated that they are willing to make.
- *SLA Management* focuses on two well differentiated phases: the negotiation of the contract and the monitoring of its fulfilment in run-time. Thus, SLA Management encompasses the SLA contract definition (basic schema with the QoS parameters), SLA negotiation, SLA monitoring and SLA enforcement according to defined policies. The automatic QoS negotiation is far away from being solved in a business context. Several SLA schemas and negotiations protocols have been proposed (WS-Agreement, WSLA, etc), but automatic negotiation of composed services and consequently multi-step negotiation of composed SLAs is a research topic. Finally, monitoring (QoS metrics) and resilience strategies are another important topic to support the future envisaged business environment.

The *Data Management* thematic area focuses on analysing the typical data management problems that businesses will encounter, and on producing patterns and advice on effective strategies for solving these problems. Data management is an area with many opportunities for expanding the current capabilities of distributed systems and Grid computing middleware in particular. It is an important area of research that underpins several aspects at the foundation of distributed computing and Service Oriented Infrastructures. It is concerned with the storage, access, translation and integration of data; it aims to answer questions like: *where should I store or cache my data? How should I get to my data? How can I present my data in a way that others will understand? How can I distribute my data and how can I combine data from different places?* In consultation with the BEs, we identified three major sub-themes, which reflect the needs of the BEs in the first phase of the project: (i) *Accessing distributed data sources*, (ii) *Accessing heterogeneous data sources, and* (iii) *Data transfer*.

The *Portals* thematic area focuses on Web interfaces for interacting with Service Oriented Infrastructures in general and Grid environments in particular. It is further divided in the following three thematic sub-areas:

1. *Administration*, which is concerned with portal administrative needs that includes providing Web interfaces for portal users with administrative roles or otherwise managing the experience of portal users.
2. *Data/information management*, which is concerned with data and information access or management needs identified in BE requirements documents with respect to portals.
3. *Job management*, which is concerned with the submission, the monitoring and control, and the visualisation of computational processes (i.e. "jobs").

2.3.3 Prioritisation of Common Technical Requirements

At the end of the common technical requirements elicitation stage, representatives from the pilot projects, technical experts and business consultants were brought together in order to consolidate the results of the elicitation exercise, and prepare the ground for the identification of common capabilities. They focused on analysing the business value of each of the common technical requirements in relation to the competitive positioning of the associated BE in their vertical market sector.

Then the common technical requirements were prioritised in two iterations:

Firstly, all common technical requirements in each thematic area were prioritised in accordance to the following criteria:

- *Popularity*: How many BE relate to this common technical requirement in the cluster?
- *Technical novelty*: How challenging and otherwise unavailable the solution will be? What is the probability of the solution generating new and significant intellectual property or technological innovation?
- *Business value*: What is the business case behind the requirement? Will it enable or facilitate the application of new more suitable business models that would have been otherwise impossible to take advantage of? Can a solution provide clearly identifiable means for revenue generation? Assume that the requirement is not met; would this have a substantial impact on business?

Secondly, the resulting prioritisation was normalised by dependences within and across thematic areas in accordance to their criticality because of interdependences.

- Critical interdependences: how many other highly ranked requirements depend upon it?

Criticality of interdependence is captured as a function that increments the priority of a common technical requirement proportionally to the accumulative value of the priority of other common technical requirements that depend upon it. This normalisation is necessary in order to create a comprehensive understanding of the business impact of the identified challenges in each thematic area. A table summarising the prioritisation of common technical requirements is available at the Technological Solutions part of IT-tude.com [6].

2.3.4 Common Capabilities and Design Patterns

Following the prioritisation of technical requirements, the technical expert teams of the BEinGRID programme reviewed and analysed architecture of the most relevant pilot projects in the programme (based on the requirements analysis of each BE). The architecture of each pilot was documented based on design templates offered to the BE by the technical expert teams. These templates were structured so as to facilitate the identification of the identification of commonalities in functionality and processes among the solutions of different BE. This analysis and subsequent research by the technical expert teams resulted in the common capabilities, and associated design patterns. A selection of common capabilities is presented in this book. A wider collection is elaborated in more technical terms at the Technological Solutions sections of IT-tude.com and the technical white papers that are being made available there.

Design Template

The design template given to all business pilots consisted of the three major parts:

Challenges and high-level architecture: The first part focused on a high-level snapshot of the system functionality described in the BE (Architecture Overview). This included a presentation of the *motivation* for the architectural choices, an explanation of the context within which each system will operate, a summary of the specific *challenges* addressed by the proposed solution, a *solution statement* explaining how the solution operating in the given context meets these challenges, a *functional overview* of the solution, and a statement of *outcome* summarising what has been achieved by the proposed design, what are the scope and limitations of the current solution in relation to the challenges that have been identified.

Thematically relevant components: The second part included a more detailed description of the system components that have been noted as relevant to a particular thematic area along with the explanation of a justification of their value. This included the identification of the components, the participating actors, their place into the overall system architecture, and a justification of the choices referring to specific selection criteria. The justification included a short description of the specific challenges being addressed by the component, an explanation of the relevance to the corresponding thematic area and an assessment of innovation, criticality for the overall architecture and perceived business impact of the availability or absence of the selected functionalities.

Value-adding components out the BE scope: The third part of the design template concentrated on capturing visions of functionality or components that would be ideally valuable for the BE but have not been considered either because of the size of an identifiable technological gap or because they would be of use further down the exploitation route of the BE.

The template was completed into a technical paper that was produced jointly by the BE and the technical experts teams of the corresponding thematic areas.

Additional guidance about best-practices for producing designs, using a selected BE as an exemplar case [6] assured traceability of the technical requirements and business value into the BE architecture.

These reports were further analysed by the technical expert teams in order identify common functionality, innovative suggestions and research solutions to open problems or improvements to generic solutions coming from the BE. Through a combination of architectural analysis of the BE designs and further research outside of the scope of any specific BE, the technical expert teams identified the common capabilities and design patterns presented in this book.

Common Capabilities

The purpose of the common capabilities is to capture reusable, platform-neutral service functionality that solves important business challenges across multiple vertical market sectors and can be realised on top of, or be integrated with, several Grid and Cloud Computing platforms.

Overall the BEinGRID team has produced 36 common capabilities in all thematic areas. Some of these capabilities are described in the remaining of this book. For a more extensive list of these requirements in available at the Technological Solutions part of IT-tude.com [6].

Table 2.2 Concepts for description of design patterns

Name of the pattern	We made an effort to ensure that the name of the pattern is simple and descriptive. Our intention was that names of patents provide an intuitive idea about how the pattern works. For example the "factory pattern" describes a thing which makes other things, much like a factory.
Intent also known as motivation	Explaining should one use this pattern, what is it intended to achieve. Alternative names of the pattern (e.g. as part of other thematic areas). Summary of a scenario that shows the problem addressed by the pattern. It also shows how the components which make up the pattern solve the problem. As much as possible, we have tried to use or adapt an exemplar solution that had been already identified in relation to the common capabilities and the architectural inspection of the BE. The goal is to use this concrete example to help understand the more abstract description that comes later.
Applicability	Explanation of situations when is it appropriate and useful to use the pattern.
Structure	A graphical representation of the parts which make up the pattern.
Participants	The components and actors which make up the pattern and their responsibilities.
Collaboration	How do the components work together to achieve the pattern.
Consequences	The results of using the pattern, including both good and bad.
Related patterns	Other patterns which can be used with or are related to this pattern (within a given thematic area/cluster or across clusters).

Design Patterns

The purpose of the design patterns is to explain the architectural context within which the common capabilities can be exploited in order to meet the elicited technical requirements. They also guide the production of generic components that enable the implementation of innovative solutions, which enhance SOI platforms with the functionality of the common capabilities. A detailed presentation of the design patterns is available at the Technological Solutions part of IT-tude.com [6].

2.3.5 Generic Components: Reference Implementations of Common Capabilities

Following the definition the common capabilities and development of associated design patterns as described above, the teams of cross-sector technical experts of BEinGRID analysed the gap between current SOA and Grid middleware and the common capabilities/design patterns produced in order to plan the development of reference implementations for these common capabilities.

Reference implementations took the form of either new services or reusable middleware components or enhancements of existing commercial products and services that implement the core functionality of some BEinGRID common capability. Development was performed by developer teams under a dedicated development coordinator from Atos Origin and the guidance of the leaders of each thematic area. Figure 2.5 summarises the process of producing the reference implementations and the information released with each of them.

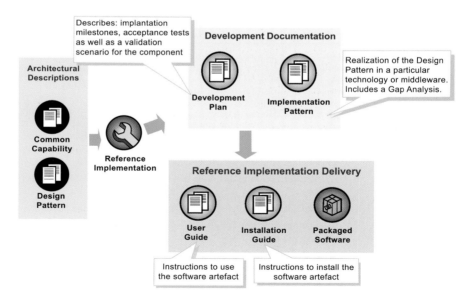

Fig. 2.5 Documentation produced with each reusable component release

After undergoing independent conformance and compliance testing, these reference implementations were assigned for validation to some of the 2nd wave of BEinGRID pilots. In some cases validation also took place within the scope of 1st wave BEinGRID pilots, where the interest from both sides was very strong and technical consultants agreed to accelerate development and validation. This has been the case for example with some of the Security and VO Management components and a business experiment offering a network-centric, federated, service hosting platform [10]. All reference implementations of Technological Solutions are accessible via the IT-tude.com web site [6]. Most of them are hosted by Gridipedia while some others are offered as enhancements of commercial products (e.g. the security enhancements of the security gateways [5] by Vordel [12] and Layer 7 Technologies [8] and the Entitlement server [4] by Axiomatics [1]) or extensions of open source research toolkits (e.g. the Web2.0 portal capabilities over the VINE toolkit [3] or SLA framework [9] over the Globus toolkit [6]).

References

1. Axiomatics corporate Web site, http://www.axiomatics.com/
2. BEinGRID—Better Business Using Grid Solutions—Eighteen Successful Case Studies Using Grid, booklet available at http://www.BEinGRID.eu/casestudies.html. See also at the BEinGRID project web site http://www.BEinGRID.eu/be.html
3. Gridipedia Technical Solution—The Vine Toolkit, http://www.gridipedia.eu/thevinetoolkit.html
4. Gridipedia—Distributed Access Control: XACML-based Authorization Service, http://www.gridipedia.eu/authorization_service.html. See also [1]
5. Gridipedia—Secure Messaging Gateway Commercial, http://www.gridipedia.eu/pep_commercial.html. See also [9] and [13]
6. Gridipedia—The European Grid Marketplace, http://www.gridipedia.eu/
7. GT4—Globus Toolkit version 4, http://www.globus.org/toolkit/
8. Layer 7 Technologies corporate Web site, http://www.layer7tech.com/
9. I. Rosenberg, A. Juan, The BEinGRID SLA framework. Report available at http://www.gridipedia.eu/slawhitepaper.html (2009)
10. D. Brossard, F. D'Andria, T. Dimitrakos, A. Gaeta (2009) Virtual Hosting Environments for Online Gaming. In: Grid and Cloud Computing: A Business Perspective on Technology and Applications. Springer, Berlin (2009, to appear)
11. The BEinGRID Project, http://www.BEinGRID.eu/
12. Vordel corporate Web site, http://www.vordel.com/

Chapter 3
Management of Virtual Organizations

Nicola Capuano, Angelo Gaeta, Matteo Gaeta,
Francesco Orciuoli, David Brossard,
and Alex Gusmini

Abstract In the Virtual Organization (VO) Management area the main challenge has been to develop policies and models for governance and lifecycle management of a business-to-business (B2B) collaboration. This work included research and development in the areas of federated identity management and semantics in addition to VO, business registries and B2B collaboration managements. The main results produced in the VO Management area include capabilities, patterns and software solutions to simplify governance and lifecycle management of B2B collaborations (VOs), and to manage applications distributed over several federated network hosts (e.g. Cloud Computing platforms).

3.1 Introduction

The activities of the VO Management area have led to the identification of Technical Requirements, Common Capabilities, Design Patterns and Software components to address the issues of governance and lifecycle management of a VO, including aspects of security and semantics in a VO.

The main challenges addressed by this area are the creation and management of a secure federated business environment among autonomous administrative domains, the separation of concerns between provision and management of application services and operational management of the VO infrastructure (e.g. separating the coordination of application execution from Resource monitoring), and the automatic discovery of available resources or services which meet a given set of functional requirements inside a VO or among different VOs.

The three key capabilities developed in this area are: (i) VO Set-up [14], this is a capability that facilitates business partner identification, and the creation and life-cycle management of a circle of trust among business partners. A competitive differentiator is that trust is aligned to consumer/provider relationships; hence

---

N. Capuano (✉)
Centro di Ricerca in Matematica Pura ed Applicata (CRMPA) c/o DIIMA, via Ponte Don Melillo, 84084 Fisciano (SA), Italy
e-mail: capuano@crmpa.unisa.it

T. Dimitrakos et al. (eds.), *Service Oriented Infrastructures*
and Cloud Service Platforms for the Enterprise,
DOI 10.1007/978-3-642-04086-3_3, © Springer-Verlag Berlin Heidelberg 2010

supporting the evolution of circle of trust to a trust network that reflects supply relationships; (ii) Application Virtualization [15], this is a composite capability that enables managing the deployment, distribution and configuration of capabilities and resources required for offering a service that is distributed over multiple hosts/cloud platforms. It offers a unifying layer for managing identity, secure service integration, SLA fulfilment and performance monitoring across multiple platforms; (iii) Automated Resource Discovery [16], a capability that improves the process of resource and service discovery in a VO by adopting semantic models and technologies.

The former two of these capabilities have been validated in a project case study demonstrating a network-centric distributed platform for scalable, collaborative online gaming [2]. The concept of a Virtual Hosting Environment that underpins this Business Experiment is an innovation that is transferable across vertical market sectors and appears to offer a generic solution for distributing services and resources in multiple Cloud Computing platforms depending on SLA requirements and offering value add by strengthening security, identity management, performance monitoring and accounting. The latter of these capabilities (automated resource discovery) has been validated in another case study focusing or sharing anti-fraud data fro roaming users within an international Group of mobile operators [5].

The rest of the chapter is devoted to introduce the main challenges of the Virtual Organization Management area, a selection of the most relevant common technical requirements, a set of common capabilities, design patterns and software components, a sample scenario showing how components interact together and how they can be collectively adopted to address a common business issue, and lastly the lessons learnt during our analysis of the case studies and some good practices identified.

The chapter is concluded with some considerations on the business adoption of the developed components.

3.2 The Main Challenges

Within the VO thematic area we have firstly tried to fix terminology and concepts. During the analysis we have taken common VO concepts from ECOLEAD [7] and TrustCoM [25] into account, in recognition of the fact that substantial basic research has already been done in the area and also that basic research on VO models and foundations is outside of the scope of the project.

Nonetheless, we think it is worth mentioning the approach for VO creation that we have selected according to the analysis of the business experiments.[1] In [6], some approaches investigated in R&D to create a VO are presented and described.

---

[1] Many case studies analyzed rely upon the service concept and WS-* family of specifications.

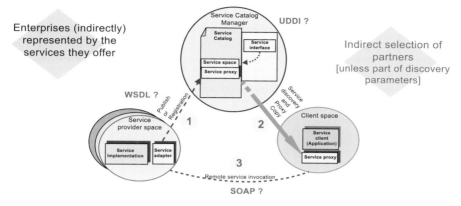

Fig. 3.1 Service Federation approach for VO

Among those, for our purpose, we are close to the so-called *Service Market based* or *Service federation* approach.

According to this approach the potential collaborative behaviour of each company is "materialized" by a set of services, and members of the VO are considered as Service Providers. The approach assumes the existence of one entity that keeps a catalogue of services where service provider companies publish their service offerings. To interact with each other, companies use standard protocols and technologies for service description, communication and data formats. Indeed, this is the approach in which the SOA, and particularly Grid based SOA like the Open Grid Services Architecture (OGSA) [9], represents a major trend in developing systems based on services.

Figure 3.1, from [6], depicts the situation.

With respect to the VO creation and management, the main problems addressed by the VOM area relate to mechanisms for federating services/resources belonging to different Service/Host Providers and facilitating the access and management to the federated services/hosts, on mechanisms allowing application and services deployment, and on mechanisms to manage the policies inside the federated group taking into account the policies of each Service/Host Provider that owns the service/resource.

Security for VO aims to improve the above mentioned mechanisms by exploiting models, standards, and specifications for secure federation. It will also take into account policies for access control and authorization mechanisms. These studies have been done in cooperation with the Security Area of the BEinGRID project.

Semantics for VO, finally, is focused on semantic annotation of services/resources and automatic services/resources discovery.

3.3 Technical Requirements

In the following sub-sections, the most relevant common technical requirements elicited from the case studies analysed are presented and described. For each one of the requirements, technical novelty[2] and business impact[3] are presented.

Before describing the selected requirements, it is worth summarising the lessons learnt during the elicitation activity. What clearly appears is that business communities are more interested in simplifying the management of heterogeneous resources in a federated business environment than in the dynamicity of the life-cycle of VOs. In particular, the business experiments analysed present common requirements related to accessing and managing, in a simple and secure way, heterogeneous distributed resources shared among the organisations participating in a collaboration. The experiments also present issues relating to resource discovery and application/service deployment.

Another key problem that emerges from the analysis is the dematerialisation of the ICT infrastructure underpinning VOs: application and ICT resource providers want to reduce or outsource the overhead of managing the distributed Service Oriented Infrastructure that underpins their Business-to-Business collaborations. It is worth mentioning that we identified business experiments—within the BEinGRID project—that present approaches with high innovation potential to address these issues.[4]

3.3.1 Secure Federation

This requirement is about the creation of a secure federated business environment among autonomous administrative domains.

The main challenges which secure federations encounter relate to the trust establishment and secure credentials distribution across multiple domains. In particular, the challenges identified relate to common federated identity issuing mechanisms, common cross organisational trust establishment mechanisms (which need to be independent of the partner-specific authentication/authorisation inside the trust realm), recognisable set of credentials, configurable solutions to support federation-related interactions.

The challenges behind this requirement are currently not addressed by traditional solutions, although research effort has been undertaken in R&D projects like Trust-CoM [25], NextGrid [19], and BREIN [1]. The business value of this requirement is very relevant for business scenarios in which actors need to establish Business-to-Business trusted relationships.

---

[2]Technical novelty indicates how challenging and otherwise unavailable a solution addressing the requirement can be.

[3]Business impact indicates if there is a concrete business case behind the requirement.

[4]It is the case, for example, of the implementation of a Virtual Hosting Environment for on-line gaming application provision. See http://www.beingrid.eu/be9.html.

3.3.2 Separation of Infrastructure Management Capabilities from Application Specific Ones

This requirement is related to the separation of concerns between the provision and management of application services (e.g. coordinating application execution, SLA monitoring) and the management of the VO infrastructure (e.g. Resource monitoring, Accounting modules, Service registry).

This requirement covers a problem faced by Application Service Providers (ASPs) that are currently responsible for managing the infrastructure and the federated hosting/execution environments. In the context of the VO life-cycle, this requirement essentially covers the set-up of the infrastructure of a VO. During the VO operational phase, common capabilities that address this requirement allow VO members to focus on managing the application level while outsourcing the administration of the underlying Service Oriented Infrastructure.

The challenges behind this requirement need an innovative solution that allows the exposure of applications in a simple, secure and manageable way without being involved in the management of the enabling infrastructure.

In terms of business impact, the requirement allows for mitigation of risks, increased flexibility (separation of responsibilities), and potential cost reduction or strategic advantage as it enables the outsourcing of infrastructure, and also handles overflow capacity and disaster recovery.

3.3.3 Automatic Resource and Service Discovery

This requirement addresses the need to discover inside a VO or among different VOs available resources/services which meet a given set of functional and/or non-functional requirements.

The challenge is to identify those services and service providers which can meet the requirements and which can reliably provide the required service, and subsequently to make a selection based upon considerations such as performance, reliability, trust, cost and quality of service.

3.4 Common Capabilities

In this section, the most relevant common capabilities are presented and described. The capabilities address recurring problems of the case studies analysed. For each capability, we describe the problem addressed, present a high-level design pattern and a sample implementation of the capability.

It is worth mentioning that during our analysis, the goal has been to abstract as much as possible the specific solutions implemented in the case studies analyzed in order to identify common capabilities that can also be reused in other contexts.

Table 3.1 summarizes the relationship between the selected technical requirements, the selected common capabilities and the software components.

Table 3.1 Selected technical requirements, capabilities and components

Common technical requirements	Common capabilities	Software components
Secure federation	VO Set-up	VO Set-up
Separation of infrastructure management capabilities from application specific ones	Creation of instances in service oriented distributed infrastructures Application Virtualization	Application Virtualization
Automatic Resource/ Service Discovery	Automatic Resource/ Service Discovery	Automatic Resource Discovery

3.4.1 VO Set Up

This capability addresses some recurring problems during the VO lifecycle, mainly in the identification and formation phases, such as partner identification, creation and management of a circle of trust among partners.

This capability is useful in typical cross-enterprise collaborative scenarios where participants (users, services, resources) have to be identified. A demand for including new participants can appear during the collaboration lifetime, and the existing participants may be dropped.

At the same time, the security of the collaboration needs to be maintained: members of a collaboration must be able to identify one another, identify messages as coming from other members of the federation, and identify the truth of claims made by other parties in the federation.

3.4.1.1 High Level Design

Figure 3.2 presents a pattern to solve the problem addressed by this capability.

The VO Set Up acts as a façade and interacts with two components: a Registry component allowing to identify potential partners of a VO and with a Federation component that is responsible for starting the creation of a circle of trust among participants.

3.4.1.2 Sample Implementation

A sample implementation of this capability is provided by the BEinGRID VO Set Up component [14].

The VO Set Up is a web service providing functionalities to support the VO lifecycle phases, and in particular the Identification and Formation phases where members of the VO have to be identified and a circle of trust among them has to

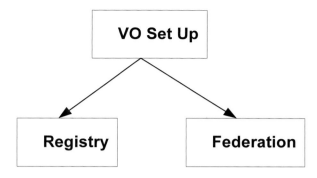

Fig. 3.2 VO Set Up

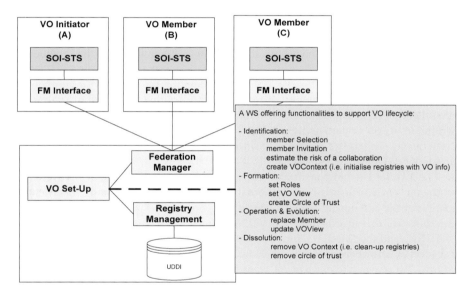

Fig. 3.3 High level architecture of the VO Set Up component

be created. The component allows the management of VO-related registries and the management of secure federation lifecycle.

A high level architecture of this component is shown in the next picture where, for completeness, are also shown the functionalities of the component divided per VO lifecycle phases. The picture shows also a possible deployment of the building blocks of the component. It is worth noting that to allow the secure federation life-cycle management, the VO Set Up interacts (via the federation manager building block) with the Security Token Service component [17] developed in BEinGRID. In the picture, the FM (Fed Manager) interface is a programmatic interface allowing to decouple the VO Set Up component from the specific SOI-STS implementation.

Each partner of a VO needs a Security Token Service (SOI-STS, which acts as an identity broker as well as a circle-of-trust enabler) and, on one partner site, the VO Set Up and its building blocks should be deployed.

This component combines VO registries and federation management in a single solution.

VO registries are built on top of UDDI standard [20] and allow the publication, discovery, and update of VO members and services. The secure federation model implemented is borrowed from the TrustCoM results [25]. The model is credential and policy-based and allows for establishment of asymmetric and binary trust relationships. The TrustCoM model has been improved and implemented, and integration with UDDI has been achieved to enable an enhanced identification phase.

The VO Set Up implements also a basic model to evaluate the risk associated with a collaboration. In its current implementation, the risk is estimated by evaluating a weighted mean of "reliability" values associated to each provider in a collaboration. The "reliability" is a metadata (implemented using the tModel structure of the UDDI standard) associated to each provider in a collaboration. The value associated to the reliability is given via feedbacks by other entities collaborating with the Provider in the past.

It is worth mentioning that the benefits of adopting the VO Set Up component is that it acts 'as a glue' among different capabilities that are required in the VO identification & formation phase. Without the adoption of this component, providers willing to trigger or participate in a VO would need to deploy and manage different components such as, for example, business registry, a service instance registry, and an Identity Management solution.

The VO Set Up component has been evaluated in the context of a concrete case study: the Virtual Hosting Environment for Distributed Online Gaming [2]. The validation of the component inside a concrete experiment has allowed us to prove the usefulness of its functionalities for VO identification and formation, the usefulness of UDDI (and the tModels defined to customise UDDI information model) as registry for VO members and VO service instances, and the process to create the circle of trust. Moreover, the experiment allowed to verify that VO registries and the federation manager could be centrally managed and configured in a coherent way via the VO Set Up component. More information on the VO Set Up evaluation can be found in [10].

3.4.2 Creation of Instances in Service Oriented Distributed Infrastructures

This capability addresses the recurring problems of service identification & creation for running and managing applications on a distributed set of resources or endpoints belonging to a Service-Oriented Infrastructure (SOI).

A common case foresees an Application Service Provider (ASP) that has to provide application capabilities to a client on the basis of an agreed contract. The ASP is a member of a VO and is aware that it can provide the application capabilities but it does not know where the application capabilities are actually deployed. It is also unaware of the status of the heterogeneous resources of the VO. For this purpose, the ASP delegates the selection of suitable hosting environments and the instantiation of the concrete services offering the required application capabilities.

3.4.2.1 High Level Design

The next picture presents a pattern to solve this problem. The client (e.g. the Application Service Provider) asks the Matchmaker for the selection of the most suitable environment. The Matchmaker performs matchmaking on the basis of application requirements and profiles of the endpoints and returns a list of suitable hosting environments. Next, the client asks for the creation of manageable service instances by invoking a high level factory, namely the Virtual Hosting Environment (VHE) factory.

The VHE factory delegates the creation of a manageable service instance to a concrete factory which, in turn, creates instances of the Management Services and of the Application Service. The endpoint references of the created instances are returned to the client.

This pattern allows the on-the-fly discovery of endpoints on which application requirements can be guaranteed and creates service instances on those endpoints. It also allows the abstraction from the specific application creation details of a particular environment. Another advantage is to decouple the application-specific logic from the management ones. To add a new family of services, the VHE factory interface has to be modified.

This pattern basically combines to the GoF Façade, Abstract Factory and Observer [11] design patterns. This pattern is also similar to the Broker Service Pattern presented in [21].

It is appropriate to apply this pattern when:

- The requestor does not know what are the suitable hosting environments where it is possible to create the instances;
- The environment within which the application operates is very dynamic, and resources are likely to register and de-register often;
- There is the need to be independent from the specific details of the creation of the family of services' instances.

3.4.2.2 Sample Implementation

Following the pattern aforementioned, the capability described in this section has been implemented in the Application Virtualization component [15].

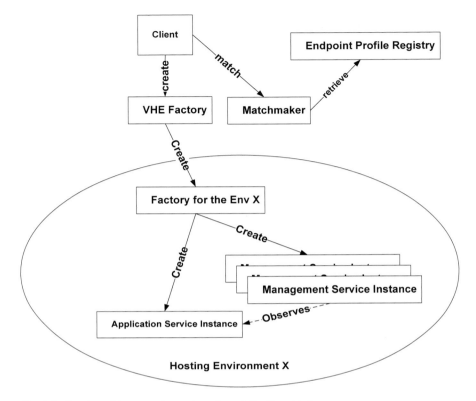

Fig. 3.4 Creation of instances in service oriented distributed infrastructures

3.4.3 Application Virtualization

This capability addresses the problem of integration and exposure of application services through a single access point (e.g. a Gateway) that is configured to manage the execution of the exposed capabilities and forward requests to them.

The capability allows an easy management of the application, taking into account policies and contracts, reducing the overhead of ASP/SP in managing the enabling infrastructure.

A common case of adoption of this capability relates to the need of exposing application capabilities (for direct usage or for composition) as network-hosted services in order to avoid direct and unmanaged access of VO resources by VO members.

3.4.3.1 High Level Design

Figure 3.5 graphically shows a pattern to address this problem.

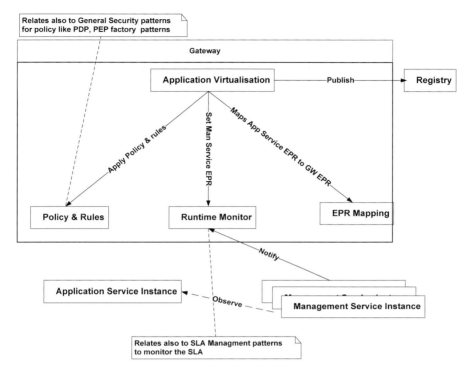

Fig. 3.5 Application Virtualization

In the picture there are notes indicating that some components relate to other BEinGRID technical areas. This means that, for instance, the Policy & Rules component can be designed and implemented according to the patterns proposed by the General Security area and the Runtime Monitoring component can be designed and implemented according to the SLA evaluation pattern proposed by the Service Level Agreement area. Interested readers can refer to the respective chapters of this book.

The Application Virtualization component follows the GoF Façade pattern and is responsible for invoking the other classes of the system in order to execute the virtualization process that consists of the following steps:

- Map the real endpoint reference of the application service instance into a virtual endpoint reference;
- Set the policies that govern who can access the application service instance and under what conditions;
- Provide the endpoint reference of the management services to a run-time monitor that is in charge of monitoring the execution (e.g. monitoring the SLA);
- Publish the virtual endpoint reference in a registry allowing other organisations/clients to discover the application service instance.

The Application Virtualization, the Runtime Monitor and the Management Service can iterate the GoF Observer pattern. Management Service Instances notify the

Runtime monitor with the updates of some parameters and the Runtime Monitor can notify violation to the Application Virtualization.

If the Application Virtualization component is also the Gateway, when a request for accessing a service arrives, the Application Virtualization can operate according the GoF Chain of Responsibility pattern and pass the request along a chain of handlers.

It is appropriate to apply this pattern when there is the need to:

- Decouple service access logic from the rest of the application
- Hide the complexities of accessing a service
- Have a single point providing common management
- Avoid direct access to resources.

3.4.3.2 Sample Implementation

The Application Virtualization component is a web service providing functionalities to create business capabilities required for the operational phase of the VO and configure infrastructural services for secure message exchange within the VO and monitoring & evaluation of the SLAs.

A high-level architecture of this component is shown in Fig. 3.6. It is possible to observe that the Policy & Rules component of Fig. 3.6 has been implemented via the triplet Secure Messaging Gateway (SOI-SMG), Authorization Service (SOI-AuthZ-PDP) and Security Token Service (SOI-STS) components of the General Security area while the Runtime Monitor of Fig. 3.6 has been implemented via the SLA monitoring and evaluation component of the SLA area. The Automatic Resource/service discovery component, instead, is presented in the following section of this chapter.

The component can be used in the VO Creation and Dissolution phases. In terms of functionalities, in fact, it allows to execute two processes.

The first one, namely the Virtualization process, consist of the following steps:

(i) Creation of services instances (business and management) on the selected hosts,
(ii) Mapping of real endpoint reference to virtual one,

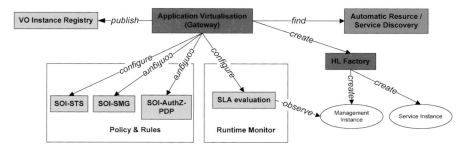

Fig. 3.6 High level architecture of the Application Virtualization component

(iii) Configuration of management services (SOI-SMG, SOI-AuthZ-PDP, SOI-STS and SLA evaluator), and
(iv) Publication of the virtual endpoint reference of the service instance into a VO Service Instance Registry. This process is executed at the end of the VO creation phase when partners that have promised to offer a service or an application in a VO need to configure their environment in order to allow secure and manageable access to that particular service instance.

The second process, namely the Graceful Shutdown, cleans up and destroys the configuration of management services, and destroys business service instances. The process consists of the following steps:

(i) Remove the service instance entries from VO Service Instance Registry,
(ii) Clean up the management services,
(iii) Clean up the Gateway (e.g. remove its internal mapping between virtual and real endpoint references), and
(iv) Destroy the business & management service instances.

3.4.4 Automatic Resource/Service Discovery

This capability addresses the recurring problem of resource and service discovery inside a VO based on a given set of functional requirements the resources need to fulfill.

The capability improves the traditional process of resource and service discovery with the adoption of semantic models and technologies.

The problem is common in several business experiments analysed for which, for instance, the scheduling and deployment of applications depend upon a number of different kinds of information, such as current workload, current application deployment, current network topology and so on.

3.4.4.1 High Level Design

Figure 3.7 shows a pattern to address this problem.

The basic idea behind the above presented design is to provide an interface to different information resources such as workload monitors, network configuration, and current application deployment information.

The participants are:

- *Resource Discovery:* interface to the subsystem. It delegates client requests to appropriate subsystem objects.
- *Run Time monitor:* A proxy for components such as Ganglia [12] and Hawkeye [18] that collect workload information from endpoints.
- *Deployment:* an interface to a database storing information about current applications deployed on endpoints.

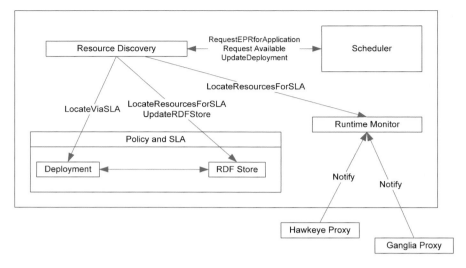

Fig. 3.7 Automatic Resource/Service discovery

- *RDF Store:* An interface to a flexible storage system based on RDF [22], which can store a variety of information without the need for a fixed schema.
- *Scheduler:* A client of the Resource Discovery subsystem.

It is appropriate to apply this pattern when the resource scheduling and allocation to endpoints depends on information from a variety of data sources, including static and dynamic information.

3.4.4.2 Sample Implementation

A sample implementation of the capability is the Automatic Resource Discovery component [16] developed in the BEinGRID project.

The Automatic Resource Discovery component is concerned with storing and retrieving Grid system information, such as Grid topology, computing and storage resources. It may also be used to store application-specific information. A key feature of the component is that the data is stored in an ontology, which supports reasoning over hierarchical data. In addition, system administrators can add deductive rules, which are automatically invoked when information is retrieved. The Automatic Resource Discovery has been designed to work with the Globus Toolkit 4 (GT4) [24] and it basically integrates a semantic layer on top of the GT4 Monitoring and Discovery System (MDS) [13].

The component is based on the RDF standard. This describes data in terms of classes, properties, and class members (instances, or objects), and informally, the data in the repository can be divided into the schema, defining classes, properties, and the relationships between classes, and instance data defining members of the classes and values for the properties.

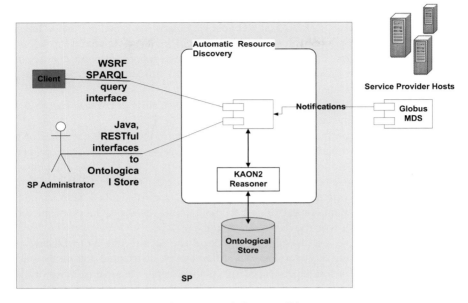

Fig. 3.8 High level architecture of the Automatic Resource Discovery component

Figure 3.8 presents a high level architecture of the component.

Resource Discovery can be done in several ways. The most popular one in a Grid environment is the adoption of the GT4 MDS. The Automatic Resource Discovery component is built on GT4 MDS and it augments the MDS index service by providing Query Service capable of executing SPARQL [23] queries.

The main advantages of using the component compared with services such as GT4 MDS is that it provides a simpler and less ad-hoc user interface, an extended information set, a common repository, and an interface for application-specific information. It reasons over an ontology rather than simple string matching of requirements against stored values.

3.5 A Sample Scenario and Integrated View of the Components

The following section presents a sample scenario involving the software components developed by the VO area. The purpose of the scenario is to show how the components can be adopted in the different phases of the VO lifecycle.[5]

The scenario defined is an application service provision scenario based on the Service Federation approach described previously in Sect. 3.2. The end-user asks for the provision of an application selecting it among a portfolio of applications that

---

[5] In accordance with the majority of the research projects that have investigated the VO paradigm, we consider the following phases of a VO lifecycle:

can be offered by an Application Service Provider (ASP). The ASP belongs to a VO Breeding Environment[6] (VBE a.k.a. Network of Enterprises), and as such it has a network of business relationships with Application Providers, Resource Providers, and Service Providers. The ASP identifies a business opportunity and decides to create a VO for the provision of the application. To this purpose, the ASP uses capabilities for VO Management.

Capabilities for VO Management can be offered in different ways. Figure 3.9 shows the two extreme cases graphically.

In the first case, on the left-hand side of the picture, there is the existence of a separate trusted third party ("VO Manager") in charge of establishing governance and rules for the federation. The VO Manager supports all the VO lifecycle phases providing services to set-up the VO, its identity management, and its infrastructure management. It is worth noting that this case differs from the Hub & Spoke model since there is not a main contractor and all the members of the VO are considered as peer (this is shown with the dashed lines in the picture).

In the second case, on the right-hand side of the picture, the VO Manager is actually a management & governance layer that can be distributed among the participant of the VO. Each independent members must deploy its own instance of the components, which must be configured to recognise the other peers of the VO.

The two approaches presented above have advantages and disadvantages: for example, the distributed management layer approach avoids the necessity for the trusted third party "VO Manager" but requires that VO members deploy VO Management services. Of course, there can also be intermediate situations where some of the functionalities (e.g. VO Registries) are offered by the VO Manager stakeholder and some are distributed among other stakeholders.

Table 3.2 presents the scenario's stakeholders, their operational and business objectives, and the business model each stakeholder follows.

The following sections details how the components can be used in the VO lifecycle phases.

- *VO Identification and formation*: it deals with identification of a goal, identification of potential partners, services, resources to achieve the goal, negotiation of agreements and policies, secure federation.
- *VO Creation*: it deals with the set-up of the VO infrastructure and creation of concrete instances of resources and services promised by the participants to achieve the goal.
- *VO Operation & Evolution*: it deals with the execution and monitoring of the tasks and business processes to achieve the goal of the VO, as well as with the management of the evolution of the collaboration (e.g. partner and service replacement, monitoring of the performance of the VO).
- *VO Dissolution*: is carried when the objectives of the VO have been fulfilled. During dissolution, the VO structure is dissolved and final operations are performed to remove all configurations, release resources of the partners, store the knowledge acquired for future collaboration.

[6]According to the ECOLEAD project, a VO Breeding Environment (VBE) represents an association or pool of organizations and their related supporting institutions that have both the potential and the will to cooperate with each other through the establishment of a "base" long-term cooperation agreement and interoperable infrastructure. When a business opportunity is identified by one member, a subset of these organizations can be selected to form a VO.

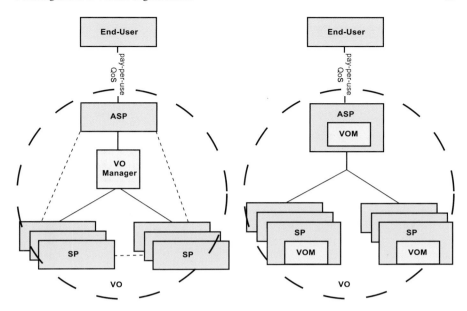

Fig. 3.9 Overview of the scenario with the VO Manager as stakeholder (*left-hand side*) and as fully distributed management layer (*right-hand side*)

3.5.1 VO Identification & Formation

The main purpose of this phase is to identify potential members of the VO, negotiate agreements, and start the secure federation process (i.e. create a circle of trust among them).

In this phase, the ASP needs to discover potential members of the VO on the basis of the capabilities they can offer. Of course, the ASP is aware of the application it needs to provide to the end-user and of the required capabilities. The ASP can query an internal catalogue containing all the partners of its VBE. Once the list of potential partners has been retrieved, ASP selects the ones it would like to collaborate with and sends an invitation to them.

The invited members, in case of acceptance, negotiate and/or sign agreements (including SLA agreements) and a policy model is defined for the VO on the basis of its objectives and of the specific members' policies. Lastly, identities of the VO members are translated into VO-wide credentials. The members are then published into a registry.

Most of the requirements of this phase can be addressed by the VO Set Up component, namely for partner identification and selection, secure federation start-up, and the publication or update of members in a VO.

Table 3.2 Scenario Stakeholders and their business opportunities

Stakeholders	Operational objective	Business objectives/values	Business model followed
Application Service Provider (ASP)	To provide application as a service on a pay-per-use model and on the basis of QoS.	To gain revenue from the provision of applications. Agility in providing its business (e.g. on-demand creation of virtual organisation to achieve the business). To save the costs of hosting all the services required for an application and of the management of infrastructure. Reduction of the total cost of ownership by outsourcing parts of the value chain. Transparent use of the Grid.	ASP as Grid User: an ASP has its existing system and wants to add some peripheral functionality. To this purpose, ASP acts as a user of the underlying Grid.
VO Manager (in case of stakeholder)	To provide capabilities to federate members (business enterprises) establishing the governance structure, rules and practices for the federation. To create and configure the underlying infrastructure for application execution.	To gain revenue from provision of federation and infrastructure management services. Cheap and fast access to Grid Computing facilities (Enabler).	Grid Enabler: it offers services to enable the collaboration between the organisations.
Service Provider (SP)	To provide VO business capabilities and resources.	To sell services and resources. Additional market by selling "component services" or "added values services", which may not be completely related to an existing business process.	Grid Service Provider: it provides service to many clients (following the classical model) however the services are of a different granularity and not necessarily consumed by the end client but also by other services.

3.5.2 VO Creation

The main purpose of this phase is the creation of business capabilities and configuration of the VO Infrastructure. This implies selection of the VO resources as well as the selection and creation of the VO service instances, and configuration of infrastructure services.

This phase can be addressed with the adoption of the Application Virtualization and Automatic Resource Discovery components.

The Application Virtualization is the component responsible for starting the process of (1) the creation of business instances that a provider has promised to offer and (2) the configuration of the VO infrastructural services. As described in the previous sections, this is done via the execution of the Virtualization Process.

The Automatic Resource Discovery component can be used to select the most suitable hosts inside the provider domain. In this scenario, the selection happens as part of the Virtualization Process.

It is worth mentioning that the approach adopted by the Application Virtualization component focuses on the separation of concerns between application provision and management of application execution. The virtualized application is exposed via a Gateway provider and access is controlled by the security services. A clear benefit of this approach is that the gateway abstracts the actual resources from the user, and enables a highly configurable and dynamic protection layer.

3.5.3 VO Operation and Evolution

In this phase, the identified partners contribute to the actual execution of the VO tasks by executing their business processes/applications. In our case, the VO has been created in order to provide an application to the end-user. Important features in this phase are the VO performance monitoring, policy enforcement (at the gateway), and exception monitoring and alerting.

When a VO member fails completely or behaves inappropriately, the VO manager may need to dynamically replace such a partner. This evolution may involve discovering new business partners, re-negotiating terms, and providing configuration information, as done in the identification and formation phase.

The operational phase is not addressed by the VO area components but can be addressed by components of other BEinGRID technical areas, such as the Security and SLA ones.

The evolution phase is partially addressed and only for the VO Member replacement. This is done in the same way as VO Identification & Formation phase via the VO Set Up.

3.5.4 VO Dissolution

The dissolution phase is carried out when the objectives of the VO have been fulfilled. During dissolution, the VO structure is dissolved and final operations are performed to remove all configurations and to release partner resources. On completion of this step, the members of a VO return to be members of a VBE.

This phase is partially executed by the VO Set Up and the Application Virtualization components.

In the case of the Application Virtualization, the "Graceful Shutdown" process is executed.

This pragmatically means the execution of a process that removes all the entries relating to the service instance to be destroyed from the Service Instance Registry, removes all the configuration information from the security services (or destroying the security service instances created for the specific instance) and SLA services. If all the clean-up steps are executed without exceptions, the actual service instance is destroyed.

After this operation, the VO members should return to being a VBE member. This is executed by the VO Set Up component that removes VO context information from the VO Member Registry.

3.6 Lessons Learnt

With respect to the analysis and support of the case studies, the VO Management area has summarised its experience in terms of identification of good practices, presentation of the main lessons learnt, and production of some recommendations for business cases relating to business-to-business collaboration.

During the analysis and support of the case studies, three different cases of collaboration and, in general, adoption of the VO paradigm have been observed. For each one of these cases, a good practice has been identified that can be followed by other business cases having similar requirements. These are:

- The case of the Grid implementation in the textile sector [3]: this is a good practice concerning the adoption of VOs for static collaboration.

  In this kind of collaboration VO members are well known and do not generally change during the lifecycle of the VO. Agreements, if present, are generally defined a priori and there is trust a priori between participants.

  The case of the digital district for textile seems to be a good practice for this kind of collaboration. The approach adopted by this business experiment is to use a Grid portal and portlet to integrate Grid technologies for resource sharing and collaborative tools.
- The case of the virtual hosting environment for online gaming [2], which identifies a good practice concerning the adoption of the VO for ad-hoc dynamic collaboration.

With ad-hoc dynamic collaboration, we refer to the case in which case the VO members have to be dynamically identified on the basis of the business goal of the VO. Of course, after the identification phase, policies and the VO agreement have to be negotiated and, generally, there is no trust a priori among the partners, so trust & identity management is a key factor for the success of this kind of collaboration.

- The case of the Grid study in oil & gas simulations [4], that appears to be a good practice concerning the re-use of already existing VO infrastructure (such as the EGEODE one [8]) for sharing of computational and data resources.

In some business cases, it may be useful to re-use existing VO and solutions already developed in the e-science community provided they fit well within the scenario. This is the case, for example, of business applications such as the finance, automotive, pharmaceutical, applications etc., that foresee as mission critical the execution of simulation, analysis of data sets and, in general, present HPC features.

In terms of lessons learnt, we have understood that, despite a strong research interest in the VO paradigm, the implementation of VO for ad-hoc dynamic collaboration (referred also as dynamic VO management) is still immature in terms of interest and adoption in e-business.

Moreover, we have observed that a current pattern (also followed by one of the BEinGRID case study) is trying to re-use existing research infrastructures in e-business mainly developed in e-science contexts. The objective is also to rely on already existing VO and solutions for VO management for computational and data resources sharing. Even if this seems to be the natural choice, re-using these existing infrastructures and solutions is suitable just for specific business cases that foresee as mission critical the execution of simulation, analysis of large dataset and, in general, present HPC-like features.

Other business cases for which, for example, ad-hoc dynamic collaboration is required or that foresee provision of services as applications should avoid this approach since, at least in its current state, the above solutions do not offer capabilities required for ad-hoc dynamic collaboration such as, for example, a rich trust management model, ability to separate among collaboration contexts and to react to contextual changes, etc. These business cases should instead consider re-using or building on top of results and findings of other projects such as TrustCoM, Akogrimo, NEXTGRID, BREIN.

Lastly, with respect to the capabilities for Virtual Organization management provided by Grid middleware, this area has observed that none of the most adopted Grid middleware offers all the capabilities required for VO management. According to our experience, this limitation has a negative impact on the adoption of VOs mainly in business domains. In addition, despite the promised "paradigm shift" (from e-science to e-business) current implementations of the most adopted middleware allows/encourages the adoption of VO paradigm mainly for computational resource sharing without paying adequate attention to the Business-to-Business collaboration aspects that underpin a commercially viable use of the VO paradigm in any business context.

In contrast, we observe the emergence of complementary technologies, such as Web Services-based Federated Identity Management or Web 2.0 based on models that place human and knowledge at the centre of the process. We believe this area to be useful to investigate emerging technologies to fulfil some lacks of the most adopted grid middleware.

A final consideration focuses on Security for VO as well as Semantics, two challenging areas proposed by the VO Management area which has had a low interest with respect to the expected one. The main reason, in our opinion, is that they have been considered by the case studies at the same time too difficult to address in the project lifetime and not mission critical. This is a big mistake mainly for security aspects that, if not addressed in early stages of development, may render difficult the process of re-engineering a prototype and, in some cases, may also prevent a potentially good solution to gain its market.

On the basis of these lessons learnt, we propose some recommendations. We essentially propose to follow one of the identified good practice, to take into consideration Grid as well as complementary technologies to address issues relating to VO management, to not underestimate the importance of security if you want to use a collaborative model and, lastly, to re-use the components developed by the BEinGRID project to gradually introduce dynamicity in collaborative scenarios.

3.7 Business Benefits

There are several business benefits associated to the results of the VO Management area.

It is worth mentioning that in general the results of the VOM area promotes an innovative model for VO that:

- Foresees an enhanced identification and formation phase, via selection of capabilities and members on the basis of SLAs, Identification of the risk associated to a collaboration, adoption of trust to mitigate risks;
- Relies on distributed trust management model;
- Fosters the adoption of Virtualization mechanisms of application and resources, via concepts such as the Virtual Hosting Environment (VHE) and B2B Gateway.

In terms of business benefits, for instance, the VO Set-up component and the capabilities implemented allows for agility in responding to new needs/requirements and improved time-to-market (by set-up of a VO when a new opportunities arises); improved trust in Business to Business interactions, and dealing with the geographical and organizational distribution of teams and computational resources.

In terms of innovation, with respect to other solutions for VO management, the model of the VO Set-up is better suited to the way enterprises thrive nowadays where new opportunities rise and fall quickly and where the environment is very prone to change. The VO Set-up allows for more flexible, business-driven interactions. Trust is established from the VO Set-up through to the security components in particular the Security Token Service.

The Application Virtualization, instead, addresses the separation of concerns between application provision and SOI operational management. The virtualized application is exposed via a Gateway and the configuration of infrastructure services (potentially provided by third parties) for managing non-functional aspects of the application is done in a transparent way for the application consumer. So, the added value is mainly in the automatic configuration of third party management services such as SLA and security. The adoption of the Gateway avoids direct access to the resources of a SP and access is controlled by the security services.

In terms of business impact, the Application Virtualization allows ASPs to expose their applications in a simple and manageable way without being involved in the management of the enabling infrastructure. This increases flexibility and allows a separation of concerns between application provision and management, and enables the transition towards a SaaS model.

In terms of exploitation opportunities, the VO Set Up component can be used in combination with components of the security area of the BEinGRID project to manage the life-cycle of circles of trust between providers targeting the Federated Identity Management market.

For the Application Virtualization, the selected strategy for this component is to be used in combination with components of the security and SLA areas of the BEinGRID project to coordinate different service execution environments to allow secure and manageable application exposure.

The idea behind this strategy is to exploit this component as a brokerage solution for different cloud providers.

3.8 Conclusion

We draw our conclusions from two perspectives.

The first perspective is the one of the VO Management area that, in our opinion, have achieved results that can be considered satisfactory. The components designed and developed cover a wide set of functionalities required to support the VO life-cycle mainly in terms of governance. We believe useful also the results in terms of patterns, capabilities and requirements that may help in improving already existing architectural solutions.

The VO Set-up is, in our opinion, a quite interesting and distinctive development with respect to other solutions for VO management, the Automatic Resource Discovery is a good improvement of the MDS of GT4 and represent a interesting work in integrating a semantic layer on top of the most adopted Grid middleware, and eventually the Application Virtualization develops a simple but effective approach to configure infrastructural services (potentially also provided by third party) for business application execution in an automatic and transparent way.

The components have been designed and developed to be modular and re-usable in several contexts. Our objective is to allow the deployment of the components into business scenarios with less effort as possible.

The second perspective for our conclusions is the one of adoption of Grid technologies for VO management in business contexts. For this perspective, the conclusion is that the traffic light is currently yellow.

Most of the visible work done so far with Grid technologies is the creation of wide research infrastructures (e-Infrastructure in EU, cyberinfrastructure in US) and so the re-use of this work also in business scenarios, as already evidenced previously in this chapter, appears to be a reasonable choice. But today this comes with a cost: mainly resource sharing aspects of the VO paradigm can be adopted by business scenarios re-using such work.

If computational resource sharing is not key and there are requirements for ad-hoc dynamic collaboration, such as agreement negotiation, trust establishment among partners, most adopted Grid middleware and solutions are still immature with the exception, of course, of some results coming from specific projects. The BEinGRID project is a source of requirements, capabilities, patterns and components, business practices, etc. that may potentially turn to green the traffic light by allowing improvement of current Grid solutions and sped-up their evolution from e-science to e-business.

References

1. Brein European IST-FP6 project, http://www.eu-brein.com/
2. Business Experiment 09 (BE09) The VHE for on line gaming application, http://www.beingrid.eu/be9.html
3. Business Experiment 13 (BE13) Virtual Laboratory for Textile, http://www.beingrid.eu/be13.html
4. Business Experiment 18 (BE18) Seismic Processing and Reservoir Simulation, http://www.beingrid.eu/be18.html
5. Business Experiment 20 (BE20) TAF GRIDS, http://www.beingrid.eu/be20tafgrids.html
6. L.M. Camarinha-Matos, I. Silveri, H. Afsarmanesh, A.I. Oliveira, Towards a framework for creation of dynamic virtual organizations, in Collaborative Networks and Their Breeding Environments (Springer, Berlin, 2005)
7. European Collaborative Networked Organisations Leadership Initiative, European IST-FP6 project, http://ecolead.vtt.fi/
8. Expanding GEOsciences on DEmand (EGEODE), http://www.egeode.org/
9. I. Foster et al., The Physiology of the Grid: An Open Grid Services Architecture for Distributed Systems Integration, http://www.globus.org/alliance/publications/papers/ogsa.pdf
10. A. Gaeta, F. Orciuoli, N. Capuano, D. Brossard, T. Dimitrakos, A service oriented architecture to support the federation lifecycle management in a secure B2B environment, in Proceeding of the Workshop Experiences on Service Oriented Infrastructure and the Grid as Foundation for the Next Generation of Business Solutions, eChallenges 2008, Stockolm, Sweden, 22–24 October 2008
11. E. Gamma, R. Helm, R. Johnson, J.M. Vlissides, Design Patterns: Elements of Reusable Object-Oriented Software (Addison-Wesley, Reading, 1995)
12. Ganglia—A Scalable Distributed Monitoring System for High-performance Computing Systems, http://ganglia.info/
13. Globus Toolkit Information Services: Monitoring & Discovery System (MDS), http://www.globus.org/toolkit/mds/
14. Gridipedia Technical Solution—VO Set-up, http://www.gridipedia.eu/vo-setup.html

15. Gridipedia Technical Solution—Application Virtualization, http://www.gridipedia.eu/application-virtualization.html
16. Gridipedia Technical Solution—Automatic Resource Discovery, http://www.gridipedia.eu/automatic-resource-discovery.html
17. Gridipedia Technical Solution—Security Token Service, http://www.gridipedia.eu/security-token-service.html
18. Hawkeye—A Monitoring and Management Tool for Distributed Systems, http://www.cs.wisc.edu/condor/hawkeye/
19. NextGrid IST-FP6 project, http://www.nextgrid.org/
20. OASIS UDDI Spec TC, http://uddi.org/pubs/uddi-v3.0.2-20041019.htm
21. O.F. Rana, D.W. Walker, Service Design Patterns for Computational Grids, 7 July 2003
22. Resource Description Framework (RDF), http://www.w3.org/RDF/
23. SPARQL Query Language for RDF, http://www.w3.org/TR/rdf-sparql-query/
24. The Globus Toolkit, http://www.globus.org/toolkit/
25. TrustCoM European IST-FP6 project, http://www.eu-trustcom.com/

Chapter 4
Aspects of General Security & Trust

**David Brossard, Theo Dimitrakos, Angelo Gaeta,
and Stéphane Mouton**

Abstract Organisations increasingly engage in business collaborations with different partners in different locations. Such enterprises want to capitalise on and offer their existing internal capabilities as services to its customers. Service-oriented architectures let them do so. SOA by definition is loosely coupled, highly granular, and often widely distributed and multi-step. They can combine internal and external services. However, exposing sensitive services through an SOA gives rise to serious security concerns. In particular, it is important to rethink identity and access management. Neither aspect is a new IT management challenge but SOA amplifies them with scale and complexity. The management of user identities, their credentials and other attributes, as well as controlling their access to the business services need to be defined, managed, controlled, and enforced. Identity silos must be bridged.

4.1 Introduction

The activities of the General Security area have led to the identification of Technical Requirements, Common Capabilities, Design Patterns and Software components to address issues of trust & security of users & applications in a distributed environment, typically regarding the privacy, confidentiality, and integrity of message exchanges between different users & services.

Key challenges come from the evolution of the way businesses interact nowadays: the work environment has become more pervasive with a mobile workforce, outsourced data centres, different engagements with customers and distributed sites. Systems are no longer monolithic: they integrate different services and clients from potentially many partners; each one with different security rules, identity stores, interfaces and regulations. Message exchanges no longer take place within the enterprise but across uncontrolled public networks. This stresses the need to secure end-to-end transactions between business partners and the customer. Companies will have to comply with their own directives and regulations as well as their partner organisations' rules and legal constraints: compliance must be monitored. In order

D. Brossard (✉)
Centre for Information Systems and Security, BT Innovate & Design, PP13D, Ground Floor,
Orion Building, Adastral Park, Martlesham Heath IP5 3RE, UK
e-mail: david.brossard@bt.com

T. Dimitrakos et al. (eds.), *Service Oriented Infrastructures
and Cloud Service Platforms for the Enterprise*,
DOI 10.1007/978-3-642-04086-3_4, © Springer-Verlag Berlin Heidelberg 2010

to enable rich & flexible scenarios, the security mechanisms put in place must support, not hinder them and must be flexible and adaptive. Different enterprises, services and customers imply multiple authorities and complex relationships regarding the ownership of resources and information across different business contexts and organisational borders. Security policies must be issued by multiple administrators and enforced over a common infrastructure. There is also a need for well-orchestrated, end-to-end Operations management that provides controlled visibility, governance of network and IT state, timely assessment of the impact of security policy violations and the availability of resources. Hence, there is an increasing interest in security observers & monitors.

One can also refer to the challenges elicited in the Virtual Organisation thematic area (see Chap. 3) to complete those already mentioned in the previous paragraph.

Five components have been developed by the General Security area over the course of the project to address these issues.

In particular, the Security Token Service (SOI-STS) (see Sect. 4.5.2) serves as an identity broker & federation manager that manages (a) an enterprise's participation in federations; (b) identity bridging between intra- and inter-enterprise identity technologies, claims, and authentication techniques; and (c) the lifecycle of identities and security attributes of users and services within that given enterprise. By federating identity brokers, a group of collaborators may create manageable circles of trust, each of them corresponding to a structurally rich trust network. The SOI-STS enables multiple administrators to control their own view of a circle-of-trust and authorized users & services. By issuing identity tokens, the SOI-STS also provides cryptographic material that can be used in secure e2e communications.

The Authorization Service (see Sect. 4.5.3) is a policy-based authorization service which takes in access control requests, evaluates them against internal policies, and returns its decision to the requestor. It grants distributed access control and combines several access control models (attribute-based, role-based, and rule-based) to produce an authorization framework suitable for highly distributed, dynamic environments. The SOI-AuthZ-PDP supports delegation which in turns enables a multiple administrative model.

The Secure Messaging Gateway (SOI-SMG) (see Sect. 4.5.4) is a policy enforcement point and an XML Security Gateway which is an appliance or software that enforces XML and Web service security policies. The SOI-SMG allows the enforcement of message and service-level policies with little or no programming. Combined with the SOI-STS or on its own, the SOI-SMG is able to analyze message flows, encrypt/decrypt, sign/validate signatures and again guarantee secure enterprise to enterprise communication. Because it is policy-based and its policy location mechanism is flexible, the SOI-SMG can allow for rich and diverse scenarios and deployments. Commercial alternatives also come with rich monitoring tools. Some of the key benefits of the SOI-SMG are that it decreases cycle time by removing security development burden from developers and coherently applying security policies across an entire enterprise.

The Security Observer (SO) (see Sect. 4.5.5) is a component that aims at monitoring security properties in a Grid environment and notifying subscribed entities when

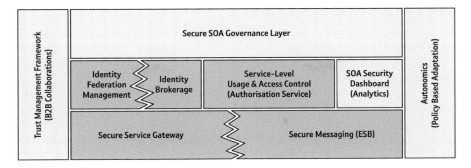

Fig. 4.1 Overview of the security capabilities required by service-oriented enterprises

something wrong has been detected on these properties. As many Grid resources are heterogeneous and deal with numerous different technologies, the associated security can become heavy to process and to maintain. In order to centralise monitoring of possible security breaches and to relieve Grid entities from security routines, the Security Observer monitors various properties and can notify any program through a standard publisher/subscriber model. The Security Observer brings a centralised and common point for security information in the grid.

Lastly, the Secure Governance Gateway (SOI-GGW) provides means to manage and configure the entire security infrastructure as well as manage the secure contextualised exposure of business services over different infrastructure profiles. It can manage the full lifecycle of policies used to configure the business and infrastructure services (see Sect. 4.5.6).

In service-oriented infrastructure, the protocols and the conditions under which service interactions occur are defined through declarative policies and agreements.

In the bottom layer of Fig. 4.1, message interceptor, message inspector, message broker and service proxy design patterns allow the enforcement of actions for service endpoints independently of the application logic. Actions will be based on sets of rules that can be specified as declarative policies that are private to a service exposure. They will specify behaviour that focuses on non-functional requirements and therefore complements the business application logic, which focuses on meeting the service's functional requirements. This is what the SOI-SMG aims at implementing.

The security components can be brought together in order to create a richer, finely adaptive solution where, from an operational perspective, the SOI-SMG acts as an integration node which delegates authentication requests to the SOI-STS, authorization requests to the SOI-AuthZ-PDP, and is coupled with the security observer to monitor a given set of parameters. Brought together, these components deliver a sturdy foundation for end-to-end WS and SOA security.

As such, the SOI-STS and the SOI-AuthZ-PDP can fulfil the functionality expressed in the second layer of Fig. 4.1: identity federation management, identity brokerage, and service-level usage and access control. This layer builds on top of and is integrated through the lower layer, that of the SOI-SMG. The SOI-SO can

implement the SOA Security Dashboard and Analytics functionality by listening to
and reporting events in the overall architecture. More on analytics can be found in
Deans et al. [4]. Lastly, the SOI-GGW sits in the top layer and is responsible for
the coherent management and configuration of the business services and supporting
infrastructure.

Overall, the expected benefits fall into two categories. Firstly, the security capabilities aforementioned help in being 'right first time'. By this, we mean that it
becomes simpler for administrators to define, apply, and monitor security mechanisms. In particular, it becomes possible to write and execute different policies
for different collaborations and keep them segregated. Therefore, services can be
exposed several times in different business contexts with different security requirements & mechanisms in place tailored to the customer's specific needs. Because
all components are programmatically manageable and customizable for different
contexts (segregation of policy execution), it is possible to differentiate policies &
services used in different collaborations with different customers. This also means
we can use multiple security providers and integrate with 3rd party security services.
More importantly, from a 'right first time' perspective, we can also assess the correctness of security enforcement via the validation of the declarative policies used
in the different security components (AuthZ, AuthC, . . .). Another consequence of
policy-based security components is regulatory compliance: this is achieved via policy coordination and their ability to be audited. The second benefit category is that
of 'cycle time': time-to-market is greatly reduced when using such an infrastructure. Using a common security infrastructure that is flexible, scalable and dynamic
reduces security management overhead as well as integration timescales of value-
adding security services. The latter can also be outsourced to specialised 3rd party
services. This lets enterprises exploit economies of scale by reusing a common security infrastructure in different collaboration contexts that they may not even need
to maintain let alone create themselves anymore.

Security work performed in BEinGRID does not cover all security aspects. Others have been dealt with in other thematic areas, in particular in the VO Management
area see Chap. 3 and the Portals Security area (see Chap. 8).

BEinGRID security activity chose to focus on end-to-end (e2e) secure communications in a distributed environment. In the following paragraph, we elicit the
challenges that arise from the SOI paradigm.

4.2 The Overall Challenge

Yesterday's systems tended to be closed and tightly coupled where security could
be dealt with efficiently by using strong identity products such as internal identity stores (e.g. LDAP, AD) and transport-layer security (e.g. SSL). These security
mechanisms do not always cater for novel SOI-oriented business models that build
on top of and call for distributed systems; multi-partner organisations (and therefore
multiple sources of identity); dynamic, adaptive, & reconfigurable architectures. For
instance, it has been observed that security failures in a distributed environment are

often the result of exploiting the fuzzy boundaries of independently robust, partial security solutions [1].

This shows current security solutions cannot be directly mapped into the SOI.

The essence of a service-oriented architecture (SOI) is the delivery of ICT infrastructure (i.e., compute, storage and network) as a set of services. We take as our starting point the three-layer model introduced in [8]: all components of the ICT infrastructure (e.g. compute resource, storage and network) can be virtualised. ICT Resource virtualization inside an enterprise has several benefits: it increases efficiency with dedicated hardware being intensively utilised across the business; usage patterns are refined to optimise resource usage. ICT virtualization is targeted at those organisations that cannot afford to own an extensive infrastructure backbone or that do not want to own such an infrastructure.

However, the offering of shared resources to different customers with different needs and expectations creates new security challenges.

Companies have to comply with their own directives and regulations as well as comply with different legislations and regulations depending on the region of operation and the client or partner organisations' rules and legal constraints. We must measure and control compliance.

The presence of multiple authorities and complex relationships regarding the ownership of resources and information across different business contexts, which span across organisational borders, mean that multiple administrators must be able to define policies about entitlements, resource usage and access.

There is a need for well-orchestrated, end-to-end Operations management that provides controlled visibility, governance of network and IT state, timely assessment of the impact of security policy violations and the availability of resources. Hence, there is an increasing interest in Service Oriented Infrastructure (SOI) governance solutions and dashboards [3, 4] showing real-time state of the infrastructure including the B2B integration points.

As the workforce becomes mobile, and the organisations increase and further integrate their collaborations and share their resources, the risks associated with the exposure of corporate information assets, services and resources increase. It becomes essential that, once threats are identified, a coordinated reaction be performed in real time to adapt usage and access policies as well as business processes across the value chain in order to mitigate risk.

The BEinGRID business experiments (see [10, 18]) presented us with the challenges aforementioned. In particular, in the Online Distributed Gaming experiment [21], there is a need for a virtual hosting environment (VHE) distributed across different enterprises and offering in-the-cloud hosting.

Another experiment focusing on eHealth where patient medical data is being exchanged requires that e2e communication be secure. It also requires that document integrity be ensured at all times. Because it is distributed across different hospitals, it also requires a distributed access control mechanism. The focus here is more on the actual data security rather than the dynamicity. It is particularly important that (1) only authorized personnel access the data; (2) the data not be tampered with; (3) the data be appropriately rendered anonymous.

Overall, the challenges can be summarised as follows:

- *Managing identities and federations in dynamic business collaborations*: How to manage the life-cycle of circles of Trust? How to enhance the structure of a circle of trust? How to coordinate a network of identity brokers in order to support the life-cycle of a Virtual Organisation? How do you contextualise identity issuance, how do you manage virtual identities and claims that are specific vary between virtual communities? How do you delegate the authority to issue credentials on one's behalf within given contexts? Part of this challenge has been addressed with the VO Management theme (see Chap. 3).
- *Security autonomics in large scale, network-centric distributed systems*: How to detect or inform about contextual changes, and how to adapt in response to contextual changes in a large-scale distributed system based on local knowledge? How to adapt the way you manage, interpret and enforce security policies in a dynamic environment?
- *Distributed access management in large*-scale decentralised systems: how to manage, reason with and enforce access policies in large-scale distributed systems? How to share policy information across domains? How to manage the confidentiality of your data and access to your applications once hosted in another's environment?
- *End-to-end security and Governance*: How to achieve end-to-end security of interactions with Grid resources? How to aggregate security services in a Grid? How to securely govern aggregated security services that are distributed over the network?

4.3 Business Motivation

In order to achieve agility of the enterprise and shorten concept-to-market timescales for new products and services, IT and communication service providers and their corporate customers alike increasingly interconnect applications and exchange data in a Services Oriented Architecture. Key security challenges come from this evolution of the way businesses interact nowadays: the work environment has become more pervasive with a mobile workforce, outsourced data centres, different business collaborations with customers and distributed sites. Systems are no longer monolithic: they integrate different services and clients from potentially many partners; each one with different security rules, identity stores, interfaces and regulations. Message exchanges no longer take place within the enterprise but across uncontrolled public networks. This stresses the need to secure end-to-end transactions between business partners and customers. Companies will have to comply with their own directives and regulations as well as their partner organizations' rules and legal constraints. Compliance must be monitored. The security mechanisms put in place must support, not hinder, such rich & flexible scenarios. They must themselves be flexible and adaptive. Different enterprises, services and customers imply multiple authorities and complex relationships regarding the ownership of resources and information across different business contexts and organizational boundaries. Security policies may be issued by multiple administrators and enforced over a common infrastructure. There is also a need for well-orchestrated, end-to-end Operations management that provide controlled visibility, governance of network and IT state, timely assessment of the impact of security policy violations and the availability of resources. Hence, there is an increasing interest in security observers & monitors.

SOA requires that administrators open ports to exchange rich application messages using a wide range of application protocols (SOAP or REST over HTTP, over SMTP, FTP, . . .). In addition, application messages must be analyzed and scanned for threats, e.g. XML threats. Traditional network security appliances do not cater for this. New mechanisms must be therefore applied at different layers. Lastly, and more importantly, the distributed nature of SOA requires a redesign of identity management systems. Bringing systems together without such a re-design creates identity silos that prevent users of one domain from freely consuming services in another. Only through well-designed federated identity management architectures can enterprises offer services across domain boundaries to users and other services it does not control or define.

The need for increasing security resulting from this new architecture is so strong that, according to Gartner (April 2009), despite the worldwide economic crisis—or possibly because of it—security aspects such as identity and authentication management (IAM) remains a critical undertaking for virtually every enterprise. Indeed, IAM alone represents a growing market which accounted for almost $3 billion in revenue for 2006 (Gartner Report, April 2009) while Infonetics Research (Content Security Appliances and Software Quarterly Worldwide Market Share and Forecasts, Feb 2009) noted that "Worldwide content security gateway revenue was up 25% from 2007 to 2008, hitting $1.9B, and will grow another 10% in 2009, [. . .]. The short and long-term opportunity for content security is strong".

Through increasing business-level visibility led by data-breach headlines, security spend continues to rise and take a growing share of overall IT spending. According to Forrester, Market Overview (April 2009), security initiatives will focus on four items: (a) protecting data, (b) streamlining costly or manually intensive tasks, (c) providing security for an evolving IT infrastructure, and (d) understanding and properly managing IT risks within a more comprehensive enterprise framework. In line with this analysis, security efficiency, with lower costs and improved service, security effectiveness, including regulatory compliance and business agility and productivity were the three main business drivers which influenced the activities of the General Security area of the BEinGRID project.

4.4 Technical Requirements

Within the scope of BEinGRID, business experiments ranged from Grid-based film processing to ship-building and from SOA-oriented hosting environments to distributed financial computing. Each environment brought its own specific sets of security needs. They were aggregated and formalised into a series of requirements as a result of which the Security Theme produced the following security requirements (see [10, 18]).

4.4.1 Primary Security Requirements

4.4.1.1 Authentication (AuthC) & Authorization (AuthZ)

This topic covers the need to authenticate and authorize service accesses in a distributed environment where there is no longer a single source of identity and where several enterprises bring their own potentially different identity stores. There is a need to be able to grant a user from one realm access to a resource of another realm. We also need to be sure where messages come from and where they are going to. Similarly, where access control could once be based on an internal corporate hierarchy, this is no longer possible in a distributed, multi-enterprise environment where there is no clear hierarchy and no agreement on roles, let alone knowledge a priori of those roles. This calls for an evolution from role-based access control to rule-based access control.

Nearly all BEs that were interested in security aspects were faced with AuthC & AuthZ issues. For instance, in FilmGrid [20] where film assets are being sent to different post-production companies, it is important to authenticate the end-user downloading and manipulating the asset to be reworked. The Online Gaming Scenario [21] presents us with an interesting AuthZ use case: users of the gaming platform will be granted access to certain game instances depending on the level of their membership. If they are gold members, they will have access to a wider range of games and to game instances that have been instantiated with a higher standard of quality.

4.4.1.2 Auditing & Assurance

This more complex environment (i.e. SOA) relies upon pieces of software distributed across the network and spanning over different organisations potentially using different technologies, rules or policies. The increase of regulations and augmentation of both complexity and frequency of usage render the management of security more and more complex. To this effect a constant validation of data as well as usage is necessary. Additionally, processes and tools that assess the good quality of the security or, lack thereof, put in place insure the sustainability of the system.

Documents crossing boundaries, users of a company accessing resources of another, and the increasing amount of cross-enterprise interactions on one hand, and the increase of laws and regulations on the other require strong auditing. This requires that all policies be traceable, auditable, assigned to a particular author, stored away for historic purposes. This in turn requires policy-driven security capabilities. They should be easy to manage and auditable.

In the financial sector, as illustrated in the Financial Portfolio Management Experiment,[1] where financial services are being offered 'in the cloud', it is important to keep track of each user's actions. It is particularly important in the light of new regulations such as Sarbanes-Oxley. These regulations have been put up to avoid massive scandals such as that of Enron in the financial sector. Companies now have to prove they can audit their systems and can trace their users' actions as well as prove they have enforced the appropriate security mechanisms. Auditing and assurance are therefore paramount. In AMONG,[2] an experiment focusing on money laundering, similar audit requirements have been expressed as we need to ensure information about banks' customers are duly protected and if released to third party only released for those purposes of an anti-money laundering investigation.

4.4.2 Distributed Systems Security

Most experiments have expressed the need to communicate over insecure network domains or public networks e.g. the Internet. As such, there is a need to manage and apply message-level security mechanisms that secure a message exchange no longer (or not only) at transport layer but also at message layer by applying end-to-end secure messaging with the possible intervention of trusted intermediaries. Such an example is the Virtual Hosting Environment for Online Gaming Scenario[3] where XML messages are being exchanged over several intermediaries and where each intermediary may be able to encrypt/decrypt and sign or verify the signature of the messages coming through.

In addition, there is a need for enforcement points that control incoming/outgoing messages and apply the relevant security. Rather than being mere firewalls, these devices need to meet the requirements of message-level security and apply the relevant security mechanisms on the fly. Such security mechanisms will include digital signatures (for integrity), XML encryption, input validation, and schema validation. This should provide trust and security through message-level enforcement.

It requires policy-driven security capabilities. They should be easy to manage, dynamically reconfigurable with zero downtime, allow for distributed policy execution, and allow for multi-author policies. The latter two points relate to the ability to cluster identical security capabilities together on the one hand and multiple administrative sources.

Requirements stem again from the Virtual Hosting Environment use case but also stem from several other experiments including those experiments exposing rich service APIs on a public network as for instance the eHealth scenario presented in BEinEIMRT [22]. The latter exposes a set of rich service interfaces to manage patient data, hospital user accounts, radiotherapy treatments. . . .

---

[1] http://www.beingrid.eu/be04.html

[2] http://www.beingrid.eu/be19among.html

[3] http://www.beingrid.eu/be9.html

Identity management in federated environments is also a key requirement. Several experiments involve a virtual organisation (see Chap. 3) with several partners that do not share a common identity store. Yet there is a need to federate and manage these different identity stores. Nearly all the experiments that evolve over a distributed system without a strong grid middleware to enable it need to federate their identity without replicating their identity stores or requiring their users to recreate user accounts in the different systems. For instance, in the Virtual Hosting Environment for Online Gaming, the Gaming platform has a user base that continually grows. This should not be reflecting in the hosting environment providers. Another example is that of TravelCRM [23], a travel agency customer relationship management system where data from different travel sites are passed through Business Intelligence tools to extract customer-meaningful data.

Lastly, access control to resources needs to happen across the entire cross-enterprise architecture. There is a need for administrators of one enterprise to be able to define access control rules on resources of another enterprise. That other enterprise must therefore be able to delegate its rights to the former administrator. This is particularly clear in the Online Gaming Scenario where the hosting partners delegate their administrative rights to the gaming partner. In FilmGrid, the owners of the assets also want to be able to delegate processing rights to the employees of the post-production companies.

4.4.3 Adaptive Enforcement

Security enforcement must no longer be a static, monolithic process. Security enforcement must take into account the context of the message to which security is being applied: where does it come from, where is it going, what time is it . . . ? Based on this contextual information, the relevant security mechanisms must be applied. This is often called context-aware enforcement.

The security architecture put forward must be capable of understanding the contents of messages and based on rules decide how to secure & process a message based on its contents. In BEinEIMRT [22], the security applied needs to depend on the information being exchanged. If we exchange sensitive patient data, then that particular information needs to be adequately secured. This is often called content-aware security.

4.4.4 Data Protection & Infrastructure Security

Although this is not directly addressed in the BEinGRID Security theme, it is important to keep in mind that the systems where data is stored must be adequately secured. In particular in the case of the medical experiment, BEinEIMRT, we need storage that can guarantee confidentiality as well as integrity of the medical data being exchanged.

The architecture put forward in the Security Theme should also take into account resilience and plan for an adaptive infrastructure capable to reconfigure itself and use new services should others fail.

4.4.5 SOA Security Governance

Efficient and comprehensive security in a distributed environment is only possible if properly and coherently managed. There is therefore a need for a management layer that can control and relate different security components together in security profiles. The governance is also critical when there is a need to drive adaptation and to understand and manage runtime events that can indicate errors, exceptions, load . . . . In the Virtual Hosting Environment scenario, administrators want to retain control of those security policies. The SOA Security Governance aspect has been jointly explored with The VO Management Theme (see Chap. 3).

4.5 Common Capabilities

4.5.1 Overview

Initial work in the Security Theme has identified the following set of low-level common capabilities (CC):

- Check validity of claims: when claims are provided by a client, there needs to be validation capability which extract such claims and run them against a set of policies to determine whether they are valid. This implies that the semantics of the claims be understood across all access control points.
- Policy engine: it is important that security components be policy-based. This allows for more flexible, dynamic scenarios to be built with highly-reconfigurable infrastructure. It also allows policies for different components to be managed centrally by a governance layer. The policies can be then be audited, stored, traced, and historically managed. There are several areas where policy engines can be used. This is described hereafter.
  - Policy-based access control: A policy-based management system allows administrators to define rules based on questions such as "Is this user allowed access to this device?" or "Are users of this organisation allowed access to this device?" and then manage them in the policy system. Policy information must be independent from components to permit policies to change and to allow the reuse of components with different policies. Policies can also be modified to suit changes in the structure of organisations or changes in the security situation, e.g. more stringent controls could be put into effect on the fly. Because of the distributed nature of the architecture put forward, one can no longer assume an access control system relying on a central hierarchy or a well-defined set of

roles. On the contrary, roles cannot be assumed or known a priori. A rule-based & policy-based access control system will be able to address this.

- Policy-based security enforcement: this takes the issue of message security enforcement as described in the following common capability and considers that efficient secure capability exposure needs to happen via a policy-based enforcement gateway. This is to address the policy requirements to do with dynamicity, the ability to update, and the ability to audit.
- Event-condition-action: this capability is about policy-based event processing. It takes into account incoming events to which it is subscribed and based on the policies it contains can produce another action or event aimed at a component to be reconfigured. This has to do with automatic adaptation and self-healing systems. This capability has been eventually grouped with the security update capability.

- Secure end-to-end communication within a federation: this capability sets the basis for an enforcement gateway able to securely expose capabilities. The secure end-to-end communication addresses the integrity and confidentiality requirements aforementioned. It fulfils the requirement that has to do with trust & security through message-level enforcement.
- Security Update: In a dynamic and heterogeneous network, the detection and the action related to a security event is a challenging task. Security events can be detected locally sent to an event hub for processing and forwarded to relevant listeners who can then adapt to these events.
- Encryption level Broker: where two entities wish to communicate on a grid they need to agree on the encryption mechanism (e.g. algorithm) and level of encryption (e.g. key length). This could be decided by a global policy, i.e. this is what you must use as a condition of joining. However, where an encryption is not determined by a global policy then two entities communicating with each other will need to negotiate the encryption that they use. This capability has been offloaded to the security gateway capability (see secure end-to-end communication).

The Security Theme has delivered five key components. They address the following key areas: identity brokerage & federation management; access control & authorization; security policy enforcement; security observer; and security governance. In the following sections we cover each of these common capabilities. For more information, please refer to [2, 8].

4.5.2 Identity Brokerage & Secure Federation Management

This area covers policy-based access control, check validity of claims, and secure end-to-end communications.

Federated identity management (FIM) originates from the need to broker identity. By identity brokerage, we mean the mechanisms that allow individuals to use the same personal identification in order to authenticate and obtain a digital identity to the networks of more than one enterprise in order to conduct transactions.

Where FIM entails managing identities across security domains, secure federation has wider objectives and stronger focus on infrastructure. To that extent secure federation can be perceived as a foundation accommodating FIM solutions. Federation solutions aim to provide interoperable service interfaces and protocols in order to enable enterprises to securely issue, sign, validate, and exchange security tokens that encapsulate claims that may include, but are not restricted to, identity and authentication-related security attributes. For more information on federations please refer to Chap. 3 as well as [6, 7].

A trust realm is an administered security space in which the source and target of a request can determine and agree whether particular sets of credentials provided by a source satisfy the relevant security policies of the target. The target may defer the trust decision to a third party (if this has been established as part of the agreement), thus including the trusted third party in the trust realm.

Message exchanges between entities in a trust realm are typically supported by services that:

- Issue, validate, and exchange security-related information such as security tokens, security assertions, credentials and security attributes (the latter can be used for policy-based access control).
- Correlate and transform such security-related information.
- Make decisions on the basis of policies that use such security information in order to determine an entity's entitlements for a given context and the way that internal authentication mechanisms map to commonly agreed security attributes (this corresponds to the CC 'check validity of claims').

4.5.2.1 Architecture

The General Security Theme has designed an identity brokerage capability (code-named SOI-STS) that meets the requirements previously elicited. This capability is based on work done during a collaboration with an innovation team from Microsoft within TrustCoM [5, 11].

The SOI-STS exposes a management interface and an operational interface. The latter is compliant with the WS-Trust specification and consists of a list of main components the most important of which are listed in Table 4.1.

At runtime, when a client requests token issuance or validation from the SOI-STS, the latter has to determine whether a token can be issued or validated in the context of the client's request.

Each federation context has an associated 'federation selector'. A federation selector is a mechanism to map a WS-Trust message (or a management operation) to an SOI-STS configuration for a federation context. In a simple case, the federation selector could contain a unique identifier such as a UUID.

From a management perspective, the SOI-STS's interface contains two parts: a set of 'core' management methods and a single 'Manage' action which dispatches management requests to dynamically selected modules. The signature of the 'Manage' method depends on the modules integrated in a given instance of the SOI-STS.

Table 4.1 The STS core components

SOI-STS database (Repository)	A database that includes configurations of SOI-STS instances for each federation context. An instance uses internal component services and policies to be used. In particular it decides which STS business logic to apply and how to control identity issue/validation process.
Federation module	A module associated with each (class of) federation context. It consists of a *federation selector* which is a scheme that allows the determination of the applicable federation context for a standard token issuance, validation, or exchange request; and information that identifies the agent that can manage this federation description (i.e. "*federation owner*").
Federation partner provider	An internal SOI-STS component service that allows the SOI-STS to retrieve information about a circle-of-trust that is identified by a unique "federation identifier". It can answer questions such as "Is BT part of the federation?" or "What organisations do I trust in this federation?" It may also apply potential constraints about a federation partner (e.g. information disclosure policy or claim validity filtering policy).
Claims provider	This internal SOI-STS component maintains associations with internal identity providers and provides a set of claims for a given "internal" identity. This will be typically used during a token issuance process. It may also apply potential constraints about a federation context and/or an "internal" identity.
Claims validity provider	An internal SOI-STS component service. It maintains associations between federation contexts, security token types, and policies that determine the validity of security claims. It is also informed of any additional constraints that apply on recognised "external" security attribute authorities (including other identity brokers) for each federation context.
Claims transformation provider	A supplementary service that applies a rules-based transformation between taxonomies of "internal" and "external" security attributes. Information disclosure policies may also be applied in order to further constrain the execution of such transformations. This auxiliary service may be called by the Claims provider or the Claims validity provider services.
AuthC scheme selector service access provider	Auxiliary service that selects the mechanism used to authenticate an entity requesting the issuance of a token and generate the associated "proof-of-possession" information.
Service access provider	A possible extension to the claims validation provider service that allows integration with an authorization service such as the one in Sect. 4.3.
Obligation policy provider	An auxiliary service that can provide "obligation policies" to be enforced by an enforcement point such as the one described in Sect. 4.4. These are further actions to be performed in order to complete a token issuance, validation or exchange request.
SOI-STS business logic	This defines a process that uses the aforementioned component services. It is executed to issue, validate or exchange a token. The STS business logic applied depends on parameters in the associated federation configuration and the content of the request.

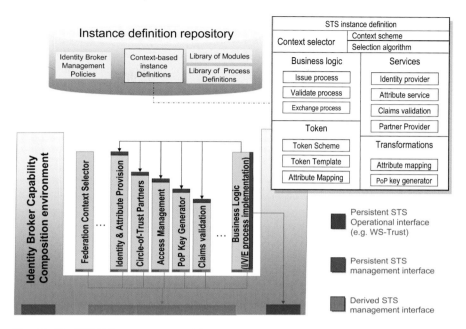

Fig. 4.2 The SOI-STS management architecture

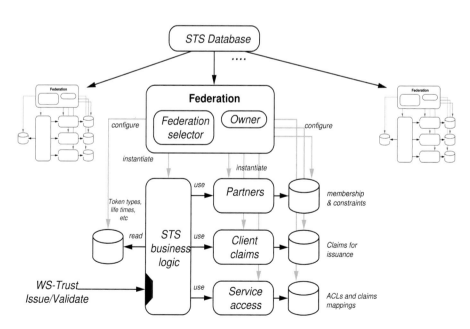

Fig. 4.3 The SOI-STS operational architecture

The flexibility of XML and SOA Web Services technology accommodates this form of dynamic composition.

The core management methods include operations for creating new federation configurations from given specifications, for enabling/disabling federation configurations or inspecting the values and meta-data of existing federation configurations. A provider management proxy function forwards provider specific management requests to the respective module. Please refer to [12] for a scenario.

4.5.2.2 Business Benefits

One of the main innovations of the selected architecture is its modularity: different modules for trust federation selection, token format, claims provisioning, authentication scheme and information disclosure policy, can be mapped by an administrator to a federation context, thereby customising the operational behaviour of the identity broker depending on the federation context.

The modularity of the architecture facilitates standards compliance and extensibility. Compliance to standards is achieved by ensuring that the core operational and management interfaces implement widely accepted standards for security token issuance, provision and exchange requests, and basic service management operations. Extensibility is facilitated by enabling the introduction of new modules implementing mission specific identity and security management models.

Different administrators can view and manage distinct configurations—hence different identity broker instances. This contextualises the SOI-STS and makes it reusable in several scenarios simultaneously. Its remote management & operational interfaces make it possible to use the SOI-STS 'in the cloud' as a value-adding (VAS) network service.

Token issuance and validation are contextualised: An identity can be issued or validated as different virtual identities (i.e. security tokens) depending on the context of the issuance request (e.g. federation context identifier, service to invoke, resource to use, action to be performed, etc).

Another innovation is that by federating such identity brokerage capabilities a group of collaborators may create manageable circles of trust, each of which corresponds to a structurally rich trust network. Each of the federated identity brokers may share the same federation context identifier (i.e. a shared state reference) and associate it with their internal view of the circle-of-trust that reflects their own trust relationships (i.e. local state). This has been further explored in the VO Management area of BEinGRID (see Chap. 3).

4.5.3 Access Control & Authorization

This area covers the following CC: check validity of claims, policy-based access control, security updates, and secure e2e communications.

Distributed Access Control (AC) and Authorization services allow the necessary decision making for enforcing groups of service-level access policies in a multi-administrative environment while ensuring regulatory compliance, accountability and security audits.

The dynamicity and level of distribution of the business models mean that one cannot rely on a set of known users (or fixed organisational structures) with access to only a set of known systems. Furthermore, access control policies need to take account of the operational context such as transactions and threat level (context-based security). This calls for a rethink of traditional models for access control and the development of new models that cater for these characteristics of the infrastructure while combining the best features from RBAC, ABAC and PBAC (please see [8] for details on each model). In the following sections we present such an authorization common capability that can cater for the requirements aforementioned.

4.5.3.1 Architecture

The General Security Theme has designed a prototype of an authorization service (SOI-AuthZ-PDP) that meets the theme's requirements (as aforementioned). Ongoing improvements are being developed in collaboration with Axiomatics [19] that aims to commercialise this prototype. Its key feature is that it implements the eXtensible Access Control Markup Language (XACML, see [14–16]).

The core elements of the information model include the policy issuer, the policy target, the policies, and the policy decision request and response.

The policy issuer is an identifiable entity that has the authority to provision access policies (including entitlements). The policy issuer may have certain entitlements about the kind of policy targets and policies that it can author and all policies issued should be signed by the corresponding policy issuer.

The policy target is the collection of variables on which a policy would apply. In the case of access management policies these may include attributes identifying some subjects, resources and actions on resources. Other environmental variables such as time, transaction context, etc., may also be taken into consideration.

Policies are collections of rules and constraints that apply on one or more policy targets. In the case of access management, policies will typically be about what actions subjects can make on resources within a scope characterised by environmental variables. Policies are combined in policy groups by means of combination algorithms that resolve conflicts by prioritising and overriding policies.

The policies fall in three categories: root, delegated, and administrative. The latter are used together in a process of validating constrained delegation of administrative authority in multi-administrative environments.

Constrained delegation validation involves looking for root policies which authorise the delegated policies in accordance with the constraints defined in the administrative policies.

Root policies are signed policies or policy sets. They are stored in a different compartment of the policy store than the delegated policies. When SOI-AuthZ-PDP loads a root policy, it will not generate a policy issuer, which must be among a collection of pre-configured trusted authorities that are established without delegation

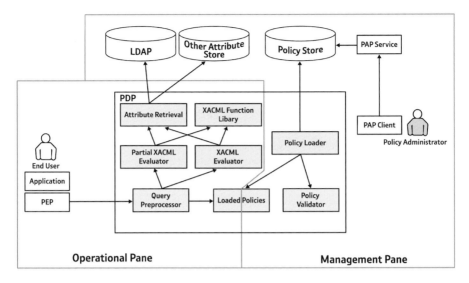

Fig. 4.4 SOI-AuthZ-PDP architecture

validation. The root policies are used to verify the authority of signed delegated policies.

Delegated policies are signed digitally by the administrative authority that issues them, i.e. the corresponding policy issuer. They are stored in a special compartment of the policy store. When SOI-AuthZ-PDP loads a delegated policy, it will use the digital signature to establish the policy's authenticity and generate a policy issuer description and associated validity constraints. The policy issuer will result in the PDP performing constrained delegation validation on the policy before it is used.

Administrative policies define the constraints that inform the administrative delegation.

Normative policy administration should happen through signed policies. The root policies define the authority of normative policy administration.

There are three interfaces in the SOI-AuthZ-PDP common capability:

- An administration interface called the Policy Administration Point (PAP) and typically exposed as a web service. It accepts XACML policies [15].
- An attribute retrieval interface that joins together adaptors to external attribute authorities.
- An operational interface: this is generally a web service implementing standard access control queries such as the XACML request profiles that have standardised bindings over SOAP and a SAML profile.

From an operational perspective, the SOI-AuthZ-PDP architecture consists of the following main components as shown in Fig. 4.4:

- Policy Enforcement Point (PEP): A requester (e.g., the end user in the figure) uses an application that contains or is deployed in a PEP. The PEP will intercept any attempted use of the application and generate an XACML request that describes the attempted access in terms of attributes of the subject, resource, action and environment. The request is sent to the PDP. The PDP will process the request and send back an XACML response, with a Permit, Not Applicable, or Deny decision, or a decision indicating an error condition, and optionally obligations. The PEP will enforce the decision and let the subject access the resource, or block the access depending on the decision. The PEP will also enforce any obligations contained in the response.

- The query pre-processor indexes the XACML query into a form which is efficient to process and generates individual queries in case the incoming request concerns multiple resources. The query pre-processor may also optimise multiple resource requests by invoking partial evaluation of XACML policies.
- The XACML evaluator evaluates the query using the XACML function modules. The XACML evaluator may retrieve additional external attributes which were not present in the incoming XACML request.

From a management perspective, the SOI-AuthZ-PDP architecture consists of the following main components (as seen in Fig. 4.4):

- A service acting as the *Policy Administration Point (PAP)*. This is the entry point for policy administration and service management. A policy administrator uses the PAP to administer the policies in the policy store. Access to the policy store is done through a PAP service which enforces invariants and access control on the policy repository. It will also perform access control on the policy store and make the required changes in the store. The PAP service will consider administration of the root policies to be a sensitive operation protected by stricter access control than administration of delegated policies.
- Attributes and Policies could be stored locally in *attribute and policy stores* or in a distributed manner (using LDAP directories for instance).
- The *policy loader* component loads policies from the *policy store*.
- The *policy validator* component is used by the *policy loader* to validate the policies syntactically, verify their digital signatures and for delegated ones, generate the policy issuer and amend administrative delegation constraints.

For a sample scenario, we invite the reader to refer to [9].

4.5.3.2 Business Benefits

One of the main requirements behind this common capability is the ability to support decentralised administration of access policies and distributed access control. This naturally occurs when the sharing of resources is spread across multiple organisations and each one wishes to keep some control over what it owns.

In distributed systems, distributed administration can reduce management costs because policy updates can be done directly and locally by the decision makers.

By introducing the concept of a policy issuer, validity constraints and by requiring policies to be signed, the authorization model ensures regulatory accountability of administration and security auditing and supports evidence gathering for regulatory compliance.

The solution supports "constrained administrative delegation". This means using policies to manage the act of creating new policies. In this context, delegation policies are about the rights to create the administrative policies. This allows the dynamic generation of delegation chains of administrative authority, which applies in complex organisations, to be constrained.

The authorization service can be contextualised and internally segmented so as to contain and segregate policy execution in different contexts (e.g. different B2B federations or transaction contexts).

This implementation is based on standards such as SAML, XACML, and WS-Security. Furthermore in collaboration with the Swedish vendor Axiomatics, the proposed policy model is being championed as a forthcoming extension (version 3.0) of XACML, the OASIS standard for access control (see [14, 15, 19]).

4.5.4 Secure Messaging Gateway

The General Security Theme designed an advanced policy enforcement point (PEP) and secure messaging gateway (SOI-SMG) that meets the requirements described above. This capability is based on prior art stemming from TrustCoM [5] and is currently being extended within BEinGRID through interactions with vendors such as Vordel and Layer 7 Technologies.

4.5.4.1 Architecture

The SOI-SMG is an appliance or software that enforces XML and Web service security policies. It allows the enforcement of message and service-level policies with little or no programming. Combined with the SOI-STS or on its own, the SOI-SMG is able to analyze message flows, encrypt/decrypt, sign/validate signatures and again guarantee secure enterprise to enterprise communication. Because it is policy-based and its policy location mechanism is flexible, the SOI-SMG can allow for rich and diverse scenarios and deployments.

In particular, it aims to deliver adaptive, extendable policy-based message level enforcement. In order to do so, while leveraging on SOA standards and patterns, different components are used. These are decomposed into Fig. 4.5.

A network of service mediation and message processing nodes (enforcement middleware): these nodes intercept each message targeted at, or originating from, a network resource or a network service endpoint. This is where service interactions are processed and service-level security policy decisions are enforced. This piece of middleware dynamically deploys a collection of message interceptors in a chain (interceptor chain) through which the message is processed prior to transmission. The interceptor chain is formed per intercepted message based on the content and context of the intercepted message as well as constraints derived from the configuration policies of the enforcement middleware. This enforcement middleware can

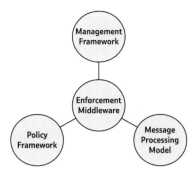

Fig. 4.5 Enforcement framework overview

also be bound to other infrastructure services (called "utility services") that undertake operations assisting the execution of enforcement actions. The binding to such utility services is explicitly declared and can be reconfigured in real time.

A policy framework that is associated with the enforcement middleware: this policy framework consists of interrelated configuration policies. The configuration policies constrain the type, execution conditions, and order of the actions enforced on the intercepted message by the selected interceptors. The configuration policies also define which external infrastructure services can be invoked by an interceptor and the conditions of such invocation.

A management framework describes the interfaces exposed by the enforcement middleware to management agents and how the management agents may interact with the system.

Finally, *two methods* are used to complete this policy enforcement model. The first method describes how management agents can create, set, update or destroy the configuration of the enforcement middleware. The second describes the enforcement middleware processes and intercepted message.

Readers should refer to [17] for details on the policy and management frameworks as well as an integrated scenario.

4.5.4.2 Business Benefits

The policy framework improves adaptation, configurability and remote management by separating concerns between enforcement policy specification, its implementation, the choice of enforcement target and the bindings between the services executing the enforcement logic in end-to-end transactions. It allows for administrators and management services to perform the following actions:

1. Dynamically update the enforcement policy and implement it on a network of heterogeneous enforcement targets;
2. Dynamically update the binding between the enforcement point and any external infrastructure services without the need to change the enforcement policy or to redeploy or update the interceptors in place;

3. Update or enhance the enforcement targets in place without necessarily having to update all configurations in place; and
4. Overcome the limitations and implicit dependencies imposed by the presence of low-level descriptions of the enforcement logic.

Furthermore the policy framework enables the provision of consistent information to clients (by means of enforceable policy obligations) and eases the propagation of policy updates between integrated services, while maintaining secrecy of the enforcement process detail on the provider's side. Readers may refer to [17] for an in-depth analysis of this benefit.

Unlike conventional firewalls or gateways, which typically define the edge of an organisation security domain, the SOI-SMG offers an instrumentation that allows integrating different enforcement targets anywhere within a network where there are XML services, or applications to protect. Typical enforcement targets include service container, application host, application gateway, network perimeter (DMZ firewall), in-cloud (e.g. router, message-broker), client back-end (e.g. add-on to service portals). Please refer to [17] for further details.

4.5.5 Security Observer

The Security Observer (SOI-SO) is a capability that has been envisioned to monitor various security properties in a Grid environment and to notify entities that compose it when certain sets of events have been recorded (e.g. a possible security flaw is detected). As many Grid resources are heterogeneous and deal with numerous different technologies, the associated security features can become heavy to process and to maintain when considered on a case by case. In order to centralise security monitoring and to relieve Grid entities from custody routines—which can be redundant—the Security Observer can be used to observe different kinds of properties all over the Grid virtual organisation and to notify any program through a standard publisher/subscriber model.

4.5.5.1 Architecture

First of all, the goal of the actual Security Observer component is to monitor several security properties on a given computing system; and then to notify some subscribers of a security event when it happens.

The architecture relies on the publish/subscribe communication model, which is implemented by two main subcomponents: the Security Monitor and the Observers. However, other relevant subcomponents bring some features to the whole application.

The Security Monitor is a daemon intended to run in background. It periodically checks security properties defined by the user. This process can be carried out through parallel threads launched by the Security Monitor itself, each thread being

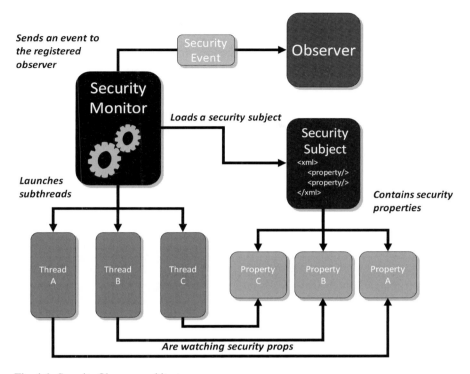

Fig. 4.6 Security Observer architecture

dedicated to a single property under consideration. A security property is represented by a Java object, built from XML statements with a set of instructions to be
executed in order to test security assumptions. The monitored security properties
can be of many types. For example, one thread can periodically execute a particular
command, observe its result and trigger an event when a specific value is found;
another thread can monitor the content of a file while a third one would periodically
check the availability of some services.

The application offers a flexible way for developers to add new modules in order
to monitor additional properties. In the beginning, the Security Monitor application
loads a bunch of security subjects, which stand for lists of security properties to
be observed, it is described in XML format. The user can thus transfer the security
properties into specific subjects based on his needs. A list of security subjects to
load has to be kept up-to-date by the user of this application.

The Security Monitor creates a security event once a singular security assump-
tion threshold is reached, and that (obviously) some anomaly is detected. This event
is sent by notification process (using Java Messaging Service, or JMS for short)
to every Observers subscribed to the Security Monitor, and roughly describes the
problem observed. Concurrently, it is possible to have any Java entity, module or
program subscribed to the Security Monitor and waiting for a notification. Such

entities are called Observers. To become so, they have to implement a specific interface provided by the Security Observer component. When an observer subscribes to a security monitor and a notification has been received, it can read the data of the security event sent and the most appropriate actions can be taken accordingly.

The action to be taken regarding the kind of event received is out of the scope of the Security Observer component. The Security Observer is thus composed of many subcomponents; amongst them the most important are the Security Monitor, the Observers, security properties, security subjects and security events.

4.5.5.2 Business Benefits

The Security Observer has been implemented as a very generic and flexible component. The core of the latter provides the mechanisms to publish and subscribe in Java, to monitor security properties defined in XML files and for Java applications to become Observers by means of a dedicated interface. Therefore, any application that wants to receive security events from the Security Monitor has just to implement the interface provided. Furthermore, the Security Monitor subcomponent is able to load external monitoring modules, given that they inherit from the generic module of the Security Observer. Hence developers can easily implement their own customised monitoring modules composed of a Java implementation and an XML descriptor, both compliant to the Security Observer architecture. Thanks to these features, the possibilities of the Security Observer component are almost infinite.

4.5.6 The SOI Governance Gateway (SOI-GGW)

The increasing amount of infrastructure services (SOI-SMG, SOI-AuthZ-PDP,...) along with all the potential states and types of configurations require adequate methods and tools for IT services governance. SOI governance is derived from corporate and IT governances. The former includes the set of processes, customs, policies, laws and institutions affecting the way in which a corporation is directed, administered or controlled. The latter focuses on the control, performance and risk of IT systems.

A SOI governance environment should offer the ability to define, administer and enforce a combination of processes, practices and tools that facilitate the management of the life-cycle of the services in the SOI as well as the life-cycle of the different policies that apply on these services.

We have developed a prototype able to manage the current security components developed within BEinGRID.

4.5.6.1 Architecture

The governance gateway is in fact an application within which are plugged in the following subcomponents.

- The Capability Instantiator (CI): this module, similar in its objective to the application virtualization component (see Chap. 3), implements the abstract factory pattern [24] to interface with the different business capabilities an enterprise may want to expose. It takes in a set of policies that describe the initial functional requirements of the business service's instantiation.
- The Federation Manager (FM): this module, similar in its objective to the VO Setup (see Chap. 3) but at a lower—partner-level—layer manages circles of trust and identity management rules for the given enterprise. In particular, it interfaces with the SOI-STS and correlates its configuration with that of the other infrastructure services. It also helps with the selection of suitable business & infrastructure policies. Eventually, it also selects a set of infrastructure services that will be used to support the exposure of business services within the given federation.
- The Virtualization Service (VS): this module implements the end-to-end virtualization process by which an enterprise exposes a business service. It first uses the capability Instantiator to create a functional instance of the business service it wishes to expose while respecting the functional requirements expressed in the instantiation policies. It also takes into account any non-functional requirements in particular those linked to quality of service and service-level agreements. Once the instantiation is complete, the VS then configures the supporting infrastructure that was selected by the FM. This includes—but is not limited to—the security services such as the SOI-STS, the SOI-AuthZ-PDP, and the SOI-SMG. The VS takes the security policies defined in the gateway registry, refines them, contextualises them, and pushes them to the relevant infrastructure services. By doing so, the VS takes the internal business instance, contextualises it, and securely exposes it within the collaboration defined by the FM.
- The Policy Store (PS): this module manages the entire set of policies and configurations defined by one or several administrators and that are to be used when configuring business and infrastructure services. In the current implementation, there are 3 levels of policies:
  - Raw policies and policy templates,
  - They are then selected within a collaboration and refined,
  - Lastly they are further refined and pushed off to the relevant infrastructure services and/or business service instance.

  The policy store enables in particular the ability to audit the entire SOI infrastructure of a given partner. It allows administrators to historically explore the policies and their refinements in order to check compliance with enterprise and legal regulations.
- Service Registries (SR): there are 3 types of service registries. Firstly, there is an internal service registry where internal, enterprise-wide, business service information is stored. In the diagram hereafter, it is labelled 'registry'. Secondly there is a public set of registries, business service white pages that can be based on the UDDI standard and which advertise services any enterprise wants to offer. Lastly, there is a constrained view of the white pages which is collaboration-specific and which only contains service information (infrastructure or business) that relates to the particular context (or collaboration) an enterprise is taking part in. The service

Fig. 4.7 Governance
gateway architecture

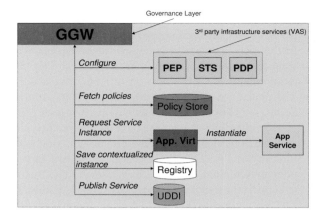

registries help in advertising the business services and their virtualised instances
along with the service level agreements and potentially other non-functional requirements such as the expected level of security.

Because each component is developed in a modular way, they can be easily replaced with another implementation thus allowing for a very flexible governance
layer. For instance the CI can be interchanged with the Application Virtualization
(see Chap. 3) with little effort.

4.5.6.2 Business Benefits

By using the SOI-GGW, administrators can easily manage business services, infrastructure service profiles, policy sets. In particular, administrators can easily control the entire state of their exposed services, monitor them, and easily adapt or
remove exposed services.

The SOI-GGW lets administrators define infrastructure service profiles. These
profiles reference the infrastructure profiles to be used as well as the processes to
configure them and how they interoperate.

The SOI-GGW lets administrators add new business services easily, define exposure policies, and instantiation processes. It gives administrators a view into each
step of the automated virtualization process.

The final innovation brought by this governance infrastructure is that it allows
managing many different types of services in various contexts. This is made possible
by the SOA used to model it and the fact that more and more industries choose to
deliver their offers as services. Please refer to [13] for further details.

4.6 Conclusion

Security is a key challenge because Grid adopters must trust the global infrastructure
and this cannot be achieved without proper security by design. This is especially dif-

ficult because of intrinsic characteristics of the Grid such as openness, heterogeneity, geographical distribution and dynamicity.

The security capabilities presented in this section address complementary aspects of security: authentication and identity management (SOI-STS); authorization and entitlement management (SOI-AuthZ-PDP); secure service exposure and security enforcement (SOI-PEP); and security attribute monitoring (SOI-SO). These security capabilities enable an enterprise to achieve the following benefits:

- To securely virtualise its applications, employee accounts, computing/information resources. This can be linked to the application virtualisation capability (see Chap. 3).
- To govern such virtualised entities: this includes defining and enforcing trust relationships enacted by the SOI-STS, security and access policies, identity schemes....
- To monitor and potentially adapt the behaviour of virtualised entities and the use of infrastructure services in response to contextual changes.
- To securely expose such assets to an open network through the SOI-SMG.
- To maintain the management of its participation in B2B collaborations. This is linked to VOM's VO Setup (see Chap. 3).
- To manage its own identities and security attributes independently from other enterprises and to.

We have also collected a series of lessons learnt. Firstly, security needs to be considered from the very inception of a grid-based or distributed system. New threats must be considered and catered for. Secondly, interoperability is essential since several specifications & different implementations are used in the distributed systems communications. In particular when it comes to web services, because Java and .NET have separate implementations for—say WS-Trust, or WSRF, it is important to thoroughly test and address these issues. Thirdly, the trust and security common capabilities (SOI-*) need to be developed in a modular way with a pluggable, extensible architecture to accommodate new security components operating with different standards (compatibility).

The Best Practices can be summarised as follows from our research:

Best Practice 1—use the software as a service (SaaS) pattern: security can be bought as a service from 'the cloud'. In order to do so, security components need to be developed as a service in an SOI fashion.

Best Practice 2—Decouple business logic from security: this avoids poor security patterns, technology lock-ins, incompatibilities, and non-extensible systems. Security should be seen as a layer which can be configured and executed independently of the business logic.

Best Practice 3—Plan for an extensible architecture: it is necessary to introduce extensibility by leveraging grid technology to develop pluggable exchangeable components.

Best Practice 4—impact: as a general rule, the introduction of security components should have minimal system performance impact.

References

1. S. Braynov, M. Jadiwala, Representation and analysis of coordinated attacks, in Proceedings of the 2003 ACM Workshop on Formal Methods in Security Engineering (ACM, New York, 2003), pp. 43–51
2. D. Brossard, T. Dimitrakos, M. Colombo, Common Capabilities for Trust & Security in Service Oriented Infrastructures, eChallenges 2008
3. P. Deans, R. Wiseman, Service oriented infrastructure: technology and standards for integration. BT Technology Journal **26**(1), 71–78 (2008)
4. P. Deans, R. Wiseman, Service-oriented infrastructure: proof of concept demonstrator. BT Technology Journal **26**(2) (2009)
5. T. Dimitrakos, TrustCoM Scientific and Technological Roadmap. Restricted TrustCoM deliverable available upon request. Contact: theo.dimitrakos@bt.com
6. T. Dimitrakos, I. Djordjevic, A note on the anatomy of federation. BT Technology Journal **23**(4), 89–106 (2005). Special Issue on Identity Management
7. T. Dimitrakos, P. Kearney, D. Golby, Towards a trust and contract management framework for dynamic virtual organisations, in eAdoption and the Knowledge Economy (IOS Press, Amsterdam, 2004)
8. T. Dimitrakos, D. Brossard, P. de Leusse, Securing business operations in an SOA. BT Technology Journal **26**(2) (2009)
9. Distributed Access Control in a Service-Oriented Infrastructure, a BEinGRID Security Whitepaper, available from Gridipedia
10. I. Djordjevic, Use case/Requirements elicitation for BE09, Grid Trust Deliverable, British Telecom, December 2006
11. C. Geuer-Pollmann, How to make a federation manageable, in Proc. of Communications and Multimedia Security, 9th IFIP TC-6 TC-11 International Conference, CMS 2005, Salzburg, Austria, 19–21 Sept. 2005
12. Federation Management & Identity Brokerage in a Service-Oriented Infrastructure, a BEinGRID Security Whitepaper, available from Gridipedia
13. Managing, Contextualizing, & Exposing Enterprise Services: a service-oriented Governance Gateway, a BEinGRID Security Whitepaper, available from Gridipedia
14. OASIS, XACML 2.0 Core: eXtensible Access Control Markup Language (XACML) Version 2.0
15. OASIS, XACML 3.0 Administration and Delegation Profile, WD 19, 10 Oct. 2007: XACML v3.0 (DRAFT) Administration and Delegation Profile Version 1.0
16. OASIS, XACML 3.0 Core Specification (DRAFT), WD 6, 18 May 2008: eXtensible Access Control Markup Language (XACML) Version 3.0 (Core Specification and Schemas)
17. Securing the Service-Oriented Infrastructure: exposing web services with XML Security Gateways, a BEinGRID Security Whitepaper, available from Gridipedia
18. L. Titkov, D1.1.1—Requirements Analysis & Technology Evaluation for General Security and Trust
19. http://www.axiomatics.com
20. http://www.beingrid.eu/be2.html
21. http://www.beingrid.eu/be9.html
22. http://www.beingrid.eu/be25.html
23. http://www.beingrid.eu/be21travelcrm.html
24. http://en.wikipedia.org/wiki/Abstract_factory_pattern

Chapter 5
Management for Service Level Agreements

**Igor Rosenberg, Antonio Conguista,
and Roland Kuebert**

Abstract Electronic services, like other general-purpose services, often need to be delivered at a guaranteed service level. Service Level Agreements (SLAs) can be used to address this by defining Quality of Service (QoS); but they usually are paper contracts. The delivery of electronic services, automatically provisioned and managed, calls for a more agile system based on dynamic SLAs: electronic contracts generated on-the-fly. The approach taken within BEinGRID to identify the barrier for wide enterprise adoption is presented. This lead to the identification of requirements, capacities and design patterns. Components were also developed to bridge the gap. Finally, after analysing the uptake of the software provided, conclusions are drawn, and recommendations are presented.

5.1 The Overall Challenge

The dynamic aspects of aggregated electronic service provisioning compel the players to expect services to be offered with a guaranteed QoS. One way to provide these guarantees is to attach the service provision with the creation of an SLA, which describes the limits of the service, and the consequences of a failure to provide it. The inability to provide SLAs is an important barrier for the Grid uptake by industry in distributed e-business environments. SLA Management encompasses the SLA contract definition (basic schema with the QoS parameters), SLA negotiation, SLA monitoring and SLA enforcement according to defined policies. The main point is to build a new layer upon the grid middleware, which is able to create negotiation mechanisms between providers and consumer of services.

The SLAs considered in this chapter are electronic contracts describing service QoS, with a short duration (hours, or days), little human intervention, and rapid deployment. This contrasts with "paper" SLAs, which are signed by lawyers, involving many human actors, have long time span, and may include elements not related to the service QoS. The electronic contracts often reference a paper SLA, called in this case the framework SLA, which defines the basic boundaries in which electronic contracts can be signed.

I. Rosenberg (✉)
Atos Origin, calle Albaracín 25, 28037 Madrid, Spain
e-mail: igor.rosenberg@atosorigin.com

T. Dimitrakos et al. (eds.), *Service Oriented Infrastructures
and Cloud Service Platforms for the Enterprise*,
DOI 10.1007/978-3-642-04086-3_5, © Springer-Verlag Berlin Heidelberg 2010

As the solution proposing dynamic SLAs was deemed quite specific and novel, the BEinGRID methodology was used for an exploration of the topic. Requirements were elicited from several Business Experiments. The initial analysis therefore started by fragmenting the whole SLA lifecycle into its separate steps. Once each challenge presented by the following Common Technical Requirements has been solved through the implementation of a component, a general framework can be built to add the SLA concepts to the whole service provisioning experience. Even more, the SLA functionality has to integrate with other capacities targeting other areas, like VO and Security. The public acceptance of SLAs for electronic services will be reached only through the integration of SLA features in a complete framework. This is what had previously been produced for the GRIA [15] and GRASP middlewares. In order to also provide a solution for a more general scenario, a solution has been implemented for the Globus Toolkit 4 middleware as BEinGRID interoperable components.

Two detailed examples are proposed in Chap. 9.

The whole SLA architecture, which should be perceived as a service management interface, must be centralised, powerful but intuitive, while being extremely verbose with monitoring information. SLAs are an elegant way of addressing variable service QoS in a competitive environment (service marketplace), but carry great possibilities of human errors inducing catastrophic economic consequences. They are still conceived, possibly rightly, as dangerous by the members of the deciding committees. Evidence is provided by the trivial and conservative SLAs offered actually (start 2009) to service customers (see for example the agreements offered by Amazon for the storage and execution services S3 [4] and EC2 [3]). In many cases these SLAs are not sufficient for the decision to rely on external providers for business-critical activities.

5.2 Technical Requirements

Providing SLAs with a service must follow the same steps as the service's own lifecycle, from creation to decommission. The SLA must be seen as the shadow of the service, in the sense that each time an event concerns the service, the event also affects the SLA. As such, the Common Technical Requirements of the SLA technical area follow the steps of the service lifecycle. The ones which have gained the most important focus in BEinGRID are: SLA Negotiation, SLA Resource Selection Optimisation and SLA Evaluation.

The reader is invited to read Chap. 9 on integration of results from different technical areas of BEinGRID, to understand how the SLA aspects are only a piece of a much greater puzzle, and how SLAs should be implemented in parallel with at least Virtual Organisation and Security features. In the following we briefly explain the main concepts.

5.2.1 SLA Negotiation

One of the most difficult problems in the SLA lifecycle is the negotiation of Service Level Agreements. The negotiation phase has to take place before any job is run, once the client has decided which provider(s) to interact with. The Service Customer and Service Provider agree on the conditions, which will be written down in the SLA document, in a bargain-like transaction: within a bargain protocol, each party tries to pull the contract to be most profitable. In the most general case, the process can last several negotiation rounds, with each party accepting more towards a compromise. Once a common ground is found, an electronic document, the SLA, is produced and signed by both parties, defining the reach of this agreement (this usually includes a service description, and its targeted QoS). This ideal setting is currently not implemented in production systems, and the negotiation protocol is cut down to accepting or rejecting the provider's offer.

The problem of reaching an agreement through consensus is difficult to solve computationally, as a typical contract negotiation in real-life would lead to a multi-phase negotiation (as mentioned above) until the customer and provider are satisfied with the consensus offer. Some of the elements which could be proposed for negotiation are the availability of resources (e.g. CPUs, memory, bandwidth, services), the usage (e.g. time, price, accessibility) and the compensations (e.g. in case of failure to comply with the agreement). Having an operator confirming each SLA negotiation on the provider side is impracticable in the case of electronic contracts. Indeed, SLAs are meant as a rapid way to establish contracts, keeping human interaction to the minimum. Automation of the contract signature is needed, bearing in mind that the document signed is expected to be (legally) binding for both parties. To achieve this, bargaining automata have to be devised: they should permit agreeing conditions within ranges, and probably with rules corresponding to business practices. This affects both the technical side with a flexible implementation, but also involves the management layers of the parties who decide which are the rules that must be followed.

5.2.1.1 Business Benefit

For negotiation, which is finalised by the SLA signature, the main challenge resides in providing a comprehensive environment for offer and demand comparison, in the legal parts as well as in the technical parts. A dedicated tool would have to eliminate the points on which an agreement has been reached, and conversely, highlight the points which remain conflicting. To provide even more insight, history of the current transaction, as well as previous ones (and their influence on the running system), should be offered, in parallel with business practices customised to the partner's profile. This means that we do not only want to take the current situation in the Consumer–Service Provider–Relation into account, but also the past behaviour of the Service Provider and its reliability in delivering services with similar properties.

The negotiation brings a better fulfilment of the user requests by fine-tuning the service description. The customised service based on the user's requirements (through the description of the SLA offer) can be delivered faster and in a more flexible manner, which will save costs and time. It also allows to adapt to the market conditions by allowing dynamic pricing: the provider can accept offers that match the market demand, and also has a chance of increasing the resource usage by proposing competitive prices. By following the WS-Agreement specification [2], it also enables better integration within the marketplace through better interoperability, which paves the way for new roles like "provider brokers", which allow the client to choose between several providers.

5.2.1.2 State of the Art and Innovation

SLA Negotiation is critical to integrate SLA functionality. Previous work has been done on other specifications (e.g. WSLA [16]), but the accepted (pre-)standard now is WS-Agreement [2], which is (June 2009) in its last steps to become a full standard. Several implementations of this protocol are available on the GRAAP-WG website [9], and most are open-source. The technical media to perform bargaining would be the WS-AgreementNegotiation protocol [1], which includes a "getQuotes" operation allowing parties to submit non-binding offers; but this specification is not yet stable.

5.2.2 SLA Optimisation of Resource Selection

This requirement essentially covers the issue of the selection of the most suitable computational resource, i.e. where it is possible to deploy and execute a service, to optimise a predefined measure of system efficiency (i.e. workload balance among the hosts, total completion time, success probability, etc.). The selection to achieve the objective is performed based on the information contained in the SLA document and on a set of information on the distributed resources.

When the SLA has been negotiated, the Service Provider has to be able to guarantee the required QoS. So, its first task before offering the SLA is to negotiate with Host Providers a suitable host (for deployment and execution of the service) or, alternatively, it has to select it among its own resources.

An objective function (function representing the goals of the allocation problem being optimised, subject to the problem's constraints) is defined with its resolution strategies. Its resolution provides the resource mapping for a given request. Depending on the considered scenario, the optimisation choices and the related resolution strategies may lead to different results. However some "basic" characteristics, that constrain the solution space (the feasible region of the problem), need to be considered independently from the particular scenario (for example SLA end date, dependency with other SLAs, pre-emptiveness, etc.).

5.2.2.1 Business Benefit

To address this requirement, matchmaking between the information of the SLA and the information on the available resources is necessary. There is also the need of associating metadata to available resources and, since the status of a resource can change dynamically, the dynamicity of the information related to the resources is a challenge to address.

To process a single SLA it is necessary to know the availability of the resources that are useful to satisfy the SLA. However, from the system optimisation point of view, a set of performance measures of the system is needed (average speed of processing an SLA, average probability of success in satisfying an SLA, etc.) in order to take into account the dynamicity of the system itself.

On the provider side, components implementing this requirement reduce the total cost of ownership of the resources. On the other side, they also allow to increase the client's confidence by assuring a better resource scheduling (less service failures). Such components are specialised in increasing the efficiency of the provider, by increasing the resource utilisation by providing better scheduling strategy (better return on investment). Another feature is the capability of ordering the requests based on business constraints like price, penalty or loyalty. These components permit the service provision to be focused on the best business value for the provider.

5.2.2.2 State of the Art and Innovation

Schedulers are designed to optimise the resource usage based on the incoming resource requests. Very few also take into account the business value of the request. This addition is a clear innovation, allowing the selection of jobs based on their value, pushing one step closer to an open marketplace. The available schedulers are only capable of handling incoming jobs. The solution proposed above considers the incoming SLAs, hence also taking into account the business value associated to the service request.

5.2.3 SLA Evaluation

Once the system is configured, the provider has to provide the resources, accept the jobs submitted, engage fault-tolerant actions, gather results, return them to the user, and most importantly monitor the execution. These actions are described in the SLA, and therefore are all compulsory. But checking the SLA in this respect can be very complicated. The user needs to be given the metrics related to the SLA, as they are agreed upon values. He/she will also be able to evaluate the performance, to take recovery/cancellation actions, or to claim compensations linked to SLA violations. For that purpose an SLA Evaluation system is needed that checks the correctness of the retrieved metrics of the SLA. Its task is mainly to provide data needed to check that the effective values offered by the system comply with the Service Level Agreement.

Based on the Monitoring information obtained through the Monitoring system mentioned in the previous paragraph, an evaluation of the SLA consists in comparing all the terms of the agreed SLA with the current situation, to discover potential violations to the agreement. This means that the data gained through the monitoring process is used to give an insight on the execution status. The evaluation of the SLA parameters can happen on internal and external data. As the SLA specifies that a given Quality of Service will be held during the duration of the contract, relative to the previously mentioned measurable elements, violations and threats can reliably be detected. Threats are defined as warning signs, in the sense that they represent SLA violation prediction. They are useful for the provider to plan preventive actions. A modified SLA can be used to configure the threat module of the evaluator component; the terms are modified to represent critical values (e.g. if a user requires a sustained minimum performance, the modified parameter describes a value higher than the violation threshold, below which the threat is sent, to notify that a dangerous situation begins).

Whenever an event is evaluated, this information has to be passed to several recipients: all the entities signing this SLA need to know the contract has been broken. The easiest way to do it is to use the WS-Notification [17] mechanism: the evaluation is a service to which potentially interested modules subscribe. When a violation is detected, each listener is notified. Some example of listeners are: recovery module, which takes recovery actions to try to circumscribe the problem or restart the job, accounting module which would need to evaluate the damage produced and issue penalties based on the SLA, and the end-user, who can decide to abort the execution.

5.2.3.1 Business Benefit

The difficulty is to be able to compare values, some coming from sensors provided by monitoring tools, others from the SLA contract itself. As the monitoring may not display all the information needed (un-cooperative providers) some estimations may have to be done, and the result can be fuzzy, with a confidence value assigned to it suggesting that the contract is violated. This could be used to take precautionary measures before a firm confirmation of violation arrives.

All parties involved in an SLA want to be able to receive information on the correct performance of the service usage. This allows changing partners when one fails to comply with its obligations, or less dramatically to claim financial compensation. This increases the client's confidence in the provider's service offering, as there is more information reaching the client. The provider's transparency is augmented. It also provides more control to the provider by having finer-tuned information on the executing services. The provider can react before failures happen, and avoid paying penalties. This reduces the provider's costs.

5.2.3.2 State of the Art and Innovation

Several monitoring tools are easily installable on a production system, like Nagios or Ganglia. More work is needed to integrate the evaluation, as this requires modelling business rules, then evaluating the functions yielding the violation status. The verification of the agreed clauses of the SLA means that the evaluation application has to have deep understanding of the agreed SLA. If this is to be run while the job is executed, WS-Notification can be envisaged as a communication media. In this domain, there may be some lessons to be learned from the TRUSTCOM European project, which proposed an evaluation functionality for WSLA [16], proposing elegant means of evaluating key performance indicators. The evaluation functionality offers another tool to discover the general status of the service contracted. The SLAs considered here are electronic contracts that are possibly signed, executed, terminated on short time frames. This contrasts with the industrial use of SLAs, which are paper documents that are measured in month or years.

5.2.4 SLA Accounting

This requirement covers the issue of calculating the price for a given service consumption taking into account the charging scheme applied. More specifically, it involves the collection of the data concerning resource consumption related to the metrics defined in the SLA contract established between the service user and the service provider and the charging of the consumer for the services provided to him by the service provider based on the stored accounting records.

The charges that will be made for various actions and the constraints placed on the client's usage of services should be contained in the SLA contract between the client and the service provider. In order to support accounting, the SLA template should contain properties such as billing period and pricing terms, according to which charges will be estimated.

At the end of each billing period, the resource usage related to a specific SLA contract will be examined and a bill will be calculated. Charges may be made for the resource reservation, as well as the cumulative usage of a metric (integral of measurement over time) or for the increase in the measurement of a metric over the billing period. Moreover, occurrences of SLA violations and thus cases of provision of services of lower quality than the agreed one in the SLA contract or even service failure can also be also recorded and reported to the SLA accounting. Hence, based on the agreed policy concerning the handling of specific penalties included in the SLA contract between the service provider and the service consumer, adjustments are made to the final bill.

What should be noted is that SLA accounting only provides the service providers with the records for the generation of the invoices to be sent to customers. Transactions involving real money should take place using the normal channels (invoices and cheques in the mail, credit card transactions on the phone, and other payment schemes). The selection of this channel can be related to the payment policy which includes payment in advance, payment during usage and payment after usage. The first one is generally regarded to be more suitable in cases when the charging of the service is fixed, the second one when charging is per service, whereas the third one when the service charge is not known in advance—without these conditions being strict however. If a client makes a payment through one of these channels, then the payment should be recorded in the account so that a correct record of the amount owed by the client is kept. Finally, the user must be able to see a detailed record of the services he used.

5.2.4.1 Business Benefit

There exist various charging schemes that could be used to form the basis of the SLA accounting scheme. The metrics included in the SLA that will be used for the charging process deal with various and heterogeneous resources. The selection and adoption of the suitable charging scheme for each business experiment based on the identified specific requirements comprises a great challenge in BEinGRID. Finally, as SLA accounting includes charging, issues of user authentication and security required are of high importance.

Accounting is a major success factor in the effective integration of grid (or service) technology in a business environment. It is the clear step between resource usage and retribution and penalties. To avoid any legal disputes, it has to be an element of the SLA, making clear under what financial conditions resources are provided, so that both consumer and provider are reassured that their interests are best served. This allows the provider to have more precise information on its billing claims. This offers more transparency, and should allow to resolve legal disputes faster, as the evidence is provided to all parties.

5.2.4.2 State of the Art and Innovation

The accounting concept has been studied extremely deeply, and many solutions exist. One existing automatic solution should be adapted to accept the Guarantees expressed in the SLAs, and seamlessly integrate SLAs as just another means of describing service offerings.

5.3 Common Capabilities

After identifying these important requirements for business users of Grid it is instructive to consider possible solutions independent of particular software. This helps to make the concepts clear before investigating the implementation details. These solutions are presented below as common capabilities.

5.3.1 SLA Negotiation Capability

The Negotiation Common Capability relates to the agreement of parties (consumer and providers) on the terms of an SLA governing the exchange of resources. The parties try to reach a deal based on a consensus after exchanging (possibly several) non-binding probes called quotes hereafter.

5.3.1.1 High Level Design

The main standards for SLAs as XML documents are WSLA [16] (now deprecated) and WS-Agreement [2]. The WS-Agreement protocol has many implementations, with more than eight referenced on the web-site of the GRAAP-WG of the OGF [9]. In general these implementations do not offer the optional getQuotes() operation, which is the difference between the original WS-Agreement protocol and its successor the WS-AgreementNegotiation protocol.

BEinGRID provided a simple SLA Negotiator component based on a plugin architecture, where the service provider only has to define the published templates and the business rules which govern the acceptance or rejection of SLA requests. The SLA Negotiator component allows to publish templates, offers the getTemplate() and createAgreement() methods. Starting from the point where the client has discovered the provider, the component allows the generation of an SLA and shares it with both parties. This component has the advantage of being very simple and targeted at only fulfilling the specification, while at the same time being easy to customise with business-driven negotiation rules described within plugins.

One important part is that in any such type of trade agreement, the provider never publishes offers. Instead, the provider waits for the client to send a pre-signed agreement. The provider expects the client's agreement to be based on a previous template. The client can reasonably know what the provider is ready to offer based on quotes that the provider has issued previously on demand. The steps are the following (see Fig. 5.1):

(1) The provider publishes a template describing the service and its possible terms, including the QoS and possible compensations in case of violation. This template leaves several fields blank or modifiable, which are meant to hold the user specific needs.
(2) The client fetches the template, and fills it in with values which describe the planned resource usage. Some terms of the template may be removed or added or changed, but may lead to a rejection in step 5. This new document, which engages neither party, called a quote request, is sent to the provider.
(3) Sending the quote request is equivalent to asking the provider to give a provisional answer, saying whether the deal could be accepted. Receiving this, the provider, based on the current resource availability and customer policies, sends back to the client a quote. This quote corresponds to values on which the provider would probably agree (but this is by no means binding), based on the client's needs expressed in the quote request.

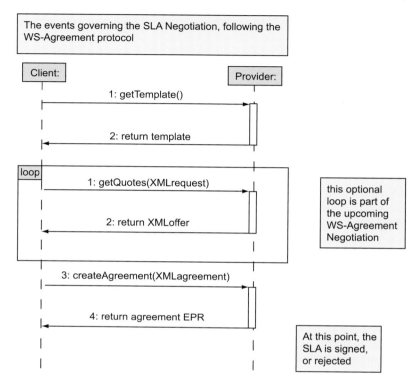

Fig. 5.1 SLA negotiation between consumer and provider

(4) The client, if satisfied with the quote, applies his/her signature to the document, and sends it back to the provider as an offer. This offer corresponds to an agreement that the user is ready to comply with. Effectively, the client is already proposing an SLA to the provider, but the provider's signature is missing.

(5) The provider, receiving the offer, is free to reject or accept it. In the latter case, the offer becomes an SLA officially signed by both parties, and starts to be a valid legal document.

Requesting the template is only done once for every type of service the provider offers. The quotes exchange (steps 2 and 3) can be repeated any number of times. The prospective service client (possibly automated) can tune the terms in the quote request until the provider's quote is in line with what the client is ready to accept. The last step for the client, step 4, requesting the real SLA, has a Boolean answer: the SLA is either accepted or rejected by the provider. In the latter case, the user can go back to asking for a quote, as the provider might have changed his conditions. This procedure allows the provider to have the final decision in accepting or declining the contract (this is crucial to avoid pre-reserving resources which may not be attributed if the client disengages).

The steps 2–3 are the core part of the negotiation, as each party can pull the deal in any direction. The parties may freely modify the different terms: lower fees, lower QoS, longer time slots, fewer resource needs, lower compensations, etc. The step 3 is the characteristic of the WSAgreementNegotiation specification [1], which by the exchange of non-binding offers, allow both parties to hint at what an acceptable offer would be.

Once a contract has been signed and agreed, the necessity of changing it could be envisaged (re-negotiation). In that case, the same quote framework could be used. The main difference comes from the already allocated resources, and the existence of a first contract to modify. Technical difficulties appear on the provider side in this case, but the framework presented to the parties to sign a new contract stays basically the same. The first call, getTemplate(), would contain as parameter the previous contract, to signify that it should be renegotiated. The procedure then stays the same as presented above.

5.3.1.2 Implementation

The BEinGRID component developed to address this capacity is the SLA Negotiator. It is targeted at the Globus Toolkit 4 middleware [7]. It is bundled with a template repository which allows to manage the SLA templates from the provider side. The component does no implement any agreement acceptation logic. Instead, it looks for available plugins and relies on those to decide of the behaviour when an agreement request is received.

More information on the SLA Negotiator can be found in the Gridipedia [8]. The Negotiation issue has now been addressed quite in depth and should be ready for industrial uptake. This can be seen by the high number of available implementations. Nonetheless, some work is still ongoing on rules defining how the client and provider should bargain (for example, by flagging which are the terms that should no longer be modified). The current state-of-the-art, for example by using the SLA Negotiator, is sufficient to allow deployment of the SLA Negotiation capability.

5.3.1.3 Example of Use

Negotiation is a crucial step to SLA-based resource exchange. The proposed capacity is adapted to SLAs which are regularly agreed, and which rely on a higher-level SLA declaring the legal validity of the electronic SLAs, and the acceptable bounds for their signature. This concept should be usable in any situation where a client requests new resources from a provider. Three BEs, which relied on the Globus Toolkit 4 middleware [7], and needed to generate an SLA, explicitly validated the component produced by BEinGRID. The basic setup is always of a prospective user discovering a provider, and then issuing a resource request based on business needs. The provider, if able to satisfy the demand, responds by creating an SLA and allocating the resources.

5.3.2 SLA Optimisation Capability

This capability allows optimisation of the execution of an application/task/service instance with an associated contract, namely the SLA, in order to maximise the probability of SLA satisfaction.

The requests, under the form of an SLA, get processed to select the best host, among all the available ones, to which the elaboration of the requests is assigned.

Obviously, the definition of a host to be 'best' depends on the state of the variables (e.g. the metrics of the SLA) in the system: available resources, resources that are needed to satisfy SLA requests and objectives to optimise. Moreover, some of the objectives are directly related to the knowledge of the requests in an SLA (completion time minimisation, cost minimisation, success probability maximisation etc.) while other objectives are related to the state of the system (for example, workload balance).

5.3.2.1 High Level Design

There are many resource schedulers available. Nonetheless, few are those that allow to optimise the resource allocation based on the value of the SLA which grants the resource.

Essentially this component is constituted by two main parts (refer to diagram below): the Optimise front-end and the GridOptimiser module. On the basis of the Publish/Subscribe patterns, the Optimisation component receives from the scheduler the job that needs to be carried out by means of an SLA offer as submitted by the scheduler user.

The Optimisation component receives from the Monitoring sub-system information about the status of the overall system. The provider is free to choose the Monitoring system that monitors the resources. Ganglia [10] and Nagios [11] are good elections. The Adapter Design Pattern permits the Optimiser Component to obtain information about the whole Grid resources from the GRID Component, while the Monitoring Component gives information about the already existing schedules and therefore about the effective availability of each resource in the grid.

All this information is processed to evaluate which optimising algorithm is more suitable and efficient to solve the scheduling problem. Once the problem is solved by the GridOptimiser component, the response is notified to the scheduler.

5.3.2.2 Implementation

Two implementations of the SLA optimisation capability were produced by BEin-GRID. The first version, designed for the GRASP middleware, was dropped after its final version was implemented due to the failure to produce reusable and correct code. Another version was produced from scratch for the GT4 middleware. This

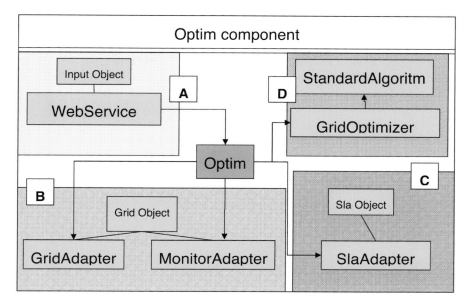

Fig. 5.2 High level architecture of the Optimisation Component

version has been published and is available as freeware binaries. It takes as input the SLA to optimise, and returns the resource allocation.

The different monitors provide information on resource utilisation. On the basis of this data, an optimisation algorithm is applied to find a solution to the allocation problem. There are currently 3 algorithms, implemented depending on the maximum resource inactivity time, free CPU, or CPU speed.

These algorithms cover the major optimisation objectives in a typical scenario. However to extend the capabilities and optimisation criteria it is necessary to increase the number of optimisation algorithms. In this way a broader set of business scenarios can be covered by applying the most suitable algorithm.

5.3.2.3 Example of Use

A scenario can be sketched from the online-gaming community: a game provider requires resources from resource providers, which must internally make sure that their resources are utilised at maximum capacity. Such a scenario was investigated in BEinGRID (see BE09 [5]), in which the resources were ordered by QoS, and the first available resource was selected. The component aimed at providing a better choice, selecting the resource based on business criteria. The component is also used in the SLA framework scenario, where it is tightly integrated to the negotiation component, offering a negotiation strategy based on the resource utilisation.

5.3.3 SLA Evaluation and Monitoring

This capability consists in monitoring the execution of a service instance (sometimes called Application, Task or Job), and evaluating if the associated contract (the SLA) is satisfied. This is needed both by provider and user. The SLA defines what are the expected performances offered by the service, and these must be retrieved from the system and compared against the values promised in the contract.

During execution, if some parameter associated to a QoS parameter of the SLA reaches a first defined threshold, a violation threat is detected and recovery actions are taken in order to guarantee the SLA, or minimise the consequences of an effective breach of the contract. If the SLO reached the second defined threshold, then a violation is detected, and is reported. Among the recovery actions, we can foresee re-allocation of application tasks on available resources, acquisition of additional resources, etc.

5.3.3.1 High Level Design

There exist many solutions which allow to monitor a system based on Key Performance Indicators. But most of these systems expect the service creation to be fairly static, in the sense that a new SLA is not often signed. In our case, to the contrary, an SLA creation is a standard event which automatically triggers a chain of events. The existing tools require human intervention for this, when we recommend (and provide means to) the seamless integration of such a functionality. The SLA Evaluation and Monitoring capability works according to the following flow:

(1) Mapping of high level application-specific business objectives (negotiated between client and provider and formalised into an SLA offer document) into low-level infrastructural parameters (such as CPU, bandwidth, memory usage) that can be quantitatively measured (usually supported by OS probes). This mapping happens on the basis of specific schemas or rules that are defined by the actor providing the application to monitor (e.g. Application or Service provider, who is the application domain expert).
(2) After the mapping of high level objectives into low-level parameters, the full SLA is created, stored and, when the client asks for the application, the SLA is retrieved and its metrics are monitored.
(3) If some parameter associated to a QoS reaches a warning threshold, in order to avoid penalties, a violation warning, called a "threat", is generated and recovery actions are taken in order to return to normal execution values.
(4) If some parameter associated breaks the QoS violation threshold, a violation of the SLA is detected. This violation is forwarded to the subscribed parties, belonging both to the user and the provider.

The capability has to allow the separation of application monitoring from monitoring of the infrastructural resources (e.g. hosts) where the application is running.

5.3.3.2 Implementation

A component providing this capability has to

(1) Interact with components that are able to monitor the parameters of the resources ("Resource Monitoring" component).
(2) Offer an interface allowing on-the-fly modification of the alarm thresholds, and allow graphical visualisation of the metrics.
(3) Analyse monitored data on the basis of rules and generate violation threat events when threat thresholds are reached.
(4) Send threat alerts to the provider allowing him to take action to try to prevent the violation.

BEinGRID has developed three distinct versions of this component. As this capability was seen as critical to a successful use of the SLA concept, several middleware were targeted, to provide a wider acceptation of the feature. The different component versions address the distinct platforms, which did not previously provide this functionality.

The first one, targeted at the GRIA middleware, addressed the issue (4) of the previous list. Indeed, GRIA is already distributed with the capability, but the evaluated information fetched from the monitored resources was not automatically forwarded to listeners. Instead, upon detecting a violation, GRIA immediately takes a corrective action by killing an appropriate number of resources with reported usage on the respective metric in order to bring the metric down. The BEinGRID component has changed the GRIA SLA Service functionality so that in case the evaluation of monitoring reports shows a violation the appropriate listeners are notified.

The other two versions have an identical architecture, but different implementation language. One uses Java and relies on Globus Toolkit 4 [7] for its WSRF interface, while the other uses .NET and its wsrf.NET capabilities. Both rely on an underlying monitoring framework to provide the resource metrics (GANGLIA [10] is already interfaced by the BEinGRID components, while Nagios [11] would require an adapter). They fetch the monitored information, evaluate it and provide threats and violations based on rules defined in the SLA.

An integrated management interface still needs to be offered with the components, so that the raw monitoring information can be visualised, to offer complementary information to the administrators when reacting to abnormal behaviour. The underlying monitoring frameworks already offer these visualisation tools, and providing extensions to reflect the SLA information could be done.

More information on the Evaluation components is available in the Gridipedia [8]. An option to this capacity is the possibility to offer information to the user who signed the SLA informing him of the current SLA status. This poses the problem of knowing who owns and runs the component implementing this capacity. If the module is on the provider side (which is logical, as the provider owns the monitoring tools which come with the resources), will the provider be willing to offer this information to the user? If on the other hand, a third reliable party own this module, how can the connection be made with the provider's monitored values?

Fine-grained information given to the user poses the problem of trust and level of implication of the user. This can only be decided by the provider, defined by its business practices.

Another point is that the component only offers the capability related to the evaluation of the SLA. To be able to perform it, the component relies on an underlying monitoring framework, which must be deployed on the provider's resource. The monitoring features relate to the resources mentioned in the SLA. It is quite straightforward for HPC-type services (computational jobs run on clusters, where a monitoring framework has usually been deployed previously). For other cases, for example SaaS models, a special monitoring device must be implemented (or installed in the case of scientific instruments) to be able to obtain the resource's metrics. Such extensions of the monitoring facilities are usually easy to implement, as the monitoring frameworks come with interfaces for new metrics.

5.3.3.3 Example of Use

On exemplary context is presented by BE09 [5], where a component provides mechanisms for monitoring and controlling a service instance during its lifecycle. The service is offered with an SLA which describes its behaviour, and the provider makes sure that the contractual QoS is respected through this component which sends notifications when errors are reported.

5.4 Motivation

This section addresses the question "Why dynamic SLAs?" We point out the business motivation and impact of the SLA concepts of this chapter, highlighting the advantages that they bring.

An exemplary situation is to extend an existing Software-as-a-Service (SaaS) solution with SLAs. The SaaS service is provided per user request, and offers more flexible pricing models (e.g. Pay-As-You-Go: paying for the usage made of the service) than the usual yearly fees of on-premise software. As such, the common software contract allowing unlimited usage may not be the most adapted to the possible irregular use of the service. The software provider business model changes dramatically here (compare SaaS provider to traditional software provider), and the SLA facilitates the SaaS business model. The dynamic aspect allows the provider to limit the management of the service, by making automatic all parts after the publication phase of the service. The service can adapt to the client's needs, and instances can be started or decommissioned when decided by the user.

The client not having control of physical resource executing the service itself may wish to have guarantees concerning the service behaviour, and possibly compensations in case of failures. "For SaaS providers, the SLA is used to set realistic expectations for their customers," says the report Setting Expectations in SaaS [13].

"The SLA clearly defines the service level commitments established by the software provider and identifies their obligations to the customer and methods of reasonable compensation should these obligations not be met". For services considered business-critical, the client will require to have minimum guarantees for the performance. The lack of proper SLAs may present the risk of preventing the adoption of SaaS for business critical applications, and this is addressed by SLAs guaranteeing the correct service provision. Even better, dynamic SLAs can be configured to fit the client's prospects.

The SLA architecture presented by BEinGRID offers the needed software components to add the SLA management functionality to software based on services, for example SaaS solutions. This architecture allows to offer the minimal SLA lifecycle, and offers the users an augmented usage experience of the original service. It provides a new management layer for finer control. This means that the target market is huge, if considered in its entirety. Nonetheless, it must be considered that not all SaaS solutions require an SLA. It can be a provider's decision to declare openly that the service is provided as-is, and that downtime may happen. This can be seen for example in the Cloud offering of Google: the Google App Engine is provided free-of-charge, but the service cannot scale out of limited bounds, and nothing covers possible downtime or data loss. The SaaS ecosystem could become bipolar, with a separation into best-effort and SLA-backed offerings. In this case, the SLA framework would target the latter segment. It also allows providers to offer distinct service levels for different clients, with the possibility of different pricing options. The different advantages brought by the SLA capacities include:

- Improved speed and flexibility through (possibly only partially) automated negotiation.
- The service QoS can be declared, monitored and evaluated against targets: the provider has an instantaneous view of the deficiencies of its provisioning based on what each client required. The user can be given more information on the service execution (metrics of the SLA), improving user awareness hence satisfaction.
- Finer control of the execution for the provider, allowing differentiation of the users based on their SLAs, to offer varying service performance based on SLAs (possibility of adapting the service provision to reflect the SLA business value when load balancing).
- The accounting algorithms (which have not been discussed here in detail) can depend on the witnessed QoS, and new accounting mechanisms (per transaction/per volume) can be offered for finer billing strategies.

5.5 Conclusion

Some best practices and recommendations for SLAs are presented in this section, based on our experiences with the BEs of the two waves. The timescales of the experiments limited the complexity of tasks they could tackle, and this in turn limited the complexity of the scenarios that could be addressed by the business experiments. Since our analysis was based on these experiments, this limited the scope of the problems we could address, and the conclusions we could infer.

5.5.1 Lessons Learned

The lessons learned during the process of Business Experiment evaluation, then through the evaluation of the components are presented as three sections: how the SLA concept was accepted, the integration for SLA frameworks, and how renegotiation is not considered critical.

5.5.1.1 SLA Concept Penetration

The main remark that can be made relates to the acceptance of SLAs. They are seen both by the service provider and client as a practical tool to describe the QoS. The second wave BEs have well reacted to the exposure to SLA concepts, and two components have been validated within 5 BEs. Nonetheless, this success should be put in contrast with the fact that the BEs had the contractual obligation to take up components.

Two BEs used the GRIA middleware, which offers a powerful framework to handle the SLA lifecycle. In this case, the SLAs are easily used as a management facility of the middleware, and pose no difficult problem. Both BEs used SLAs in order to constrain and monitor the number of the queries exchanged between two parties. In one scenario, the SLAs are used to limit and control the number of inter-bank queries and make sure that data exchanged are not used for competition purposes. In the other, the SLAs control the data sharing performed between partner agencies belonging to a same group.

Three more BEs relied on GT4, and could validate the negotiation component. It proved a simple solution to generate SLAs and was deemed satisfactory for simple business requirements. The SLA Evaluation capability was also tested in one integrated component, presenting a scenario requiring cooperation of several components together. Unfortunately there was no complete deployment of a GT4 SLA framework, due to the late development of the component integration. The global business vision of the SLA usage cannot be presented.

The missing uptake of the whole lifecycle is surprising. One explication is that comprehensive SLA functionality is strangely not yet seen as a strong business need. SLAs are not widely accepted, and still in the "early adopters" phase. Simple approaches to add SLAs are desired. But bearing in mind the manager's perspective, which requires a strong need for control, as SLAs may involve dynamically deciding purchases, the solutions must provide a proven and reliable framework. This leads to higher costs in terms of installation and technical management of the infrastructure. Hence the decision of which level of complexity should be adopted by the framework is a decision which must be taken at a high level of the hierarchy.

The components presented and developed by BEinGRID in terms of SLAs provide sufficient functionality to handle a simple SLA situation (negotiation, optimisation, monitoring, accounting). This is enough for the needs as expressed by the BEs, but could be too limited for a more complex situation involving more parties. The development of such more complex frameworks is left to future work. The interested reader could refer to the section on SLAs of the GRIA middleware [15], but should also evaluate the development within BEinGRID of the GT4 SLA framework [12]. The SLA@SOI [14] project (started June 2008) could also lead to further development.

5.5.1.2 Need for an Integrated SLA Functionality

Deciding to attach SLAs to the service offer means adding a lot of overhead on the service provision. The different steps of the lifecycle have to be addressed; the minimal setup requires at least negotiation, evaluation and accounting. As seen with the second wave BEs, adopting a middleware already having a mature SLA layer is much easier, as in the case of GRIA. BEs asking for SLAs in an environment initially thought without SLAs had more integration problems, and then found some difficulty in justifying their choice of using SLAs. An attempt to address this concern was done in the last stages of the BEinGRID project, by offering an integrated framework for SLAs, supporting the GT4 middleware. This framework provides the basic SLA lifecycle management functionality, and can be used to offer services, with the added benefit of reported QoS.

5.5.1.3 Renegotiation Is not Clearly Needed

When seen in light of current state-of-the-art (June 2009), renegotiation is still a research topic. The GRAAP-WG (WS-Agreement discussion group) only started to discuss this topic during the last months of 2007, and has not yet produced a specification draft. The reason to provide the Re-Negotiation capability is also unclear. In the context of a marketplace of Grid providers, contracts should be established on a short-term basis to allow discarding providers as competitors push prices down. It might come that re-negotiating a contract becomes more expensive than keeping it as it is, and waiting for its expiration. Imagining an SLA for a long-term relation is difficult, as these contracts already exist, but are real legal contracts written by qualified lawyers, as they represent a high risk for the companies engaging in them. This is probably the main point: long-term contract are written down on paper, whereas short-term contracts (probably referencing long-term paper contracts) can be negotiated automatically, within a limited scope, and more importantly, people are willing to do this!

The German law specifies that re-negotiation must be treated as a new negotiation. Taking this model into account, it can also be argued that the negotiation features are quite sufficient, and that the SLA specifications are already flexible enough to allow this scheme.

5.5.2 Recommendations

The BEinGRID experience on dynamic SLAs leads to two main recommendations: if SLAs will be used, it must be decided in the very beginning, and pre-SLAs can help limit the dynamic range; another important point is to make sure that offering SLAs really makes sense for the given service.

5.5.2.1 Plan Your SLA Usage in the Early Stages

From the BEinGRID experience, it appears that generating completely dynamic contracts is too difficult, or cumbersome. The user and the provider are not ready to spend time revising SLAs each time a new negotiation takes place. This is due to the legal aspects that could change with a new type of SLA, but also due to changing pricing schemes affecting human hierarchy (authorisation for buying). So it is much easier in a real business scenario to constrain the user to only describing exactly what type of jobs are going to be submitted to the Grid: services, HPC, specific applications, etc. This specification can be done by means of a pre-SLA. This pre-SLA can bear several forms. It can be a paper document, which means it can easily be included in a legal contract. It can also be an XML document, and in this case it is easily inserted in the software stack as a constraint to the possible contracts. The ideal format is a pair {paper, XML}, which reference each other, offering the benefits of both. Provider and user must agree on this initial document, which will serve as a draft for all the SLAs signed between each party. This draft limits very precisely the conditions under which an SLA can be signed, limiting within ranges the values, and limiting the services provided to a reduced number. To produce this type of document, there is a need to know:

- The metrics, which are going to be used to measure the SLA violations.
- The type of jobs which are going to be submitted.
- The frequency of submission of new SLA requests.
- The data flows for incoming or outgoing files.

Once this pre-SLA document exists, negotiation can carry on in a very simple manner. The space of possible SLAs has been reduced drastically, and has been limited to selecting values from a range. This means the end-user GUI creation is much simpler, and the complexity presented to the End-User is reduced to the core business. It also means the provider acceptation agent can limit its intelligence, up to the point that it can be easily automated and configured by a simple configuration file. This simplification is a key benefit for all, offering easy handling of SLAs.

The final point on SLA modelling concerns the specification to comply to. WSLA is efficient in specifying Web Service-related metrics, whereas WS-Agreement is slowly seeping as the accepted standard. Some proposals combine bits of WSLA in WS-Agreements to allow the proper definition of services and their QoS, the idea being to relate the metric with the method that enables to fetch it, but also the way to evaluate it (for example the "wsla:Less" predicate, but any mathematical function could be used).

5.5.2.2 Who Wants to Sign an SLA?

Usually, a strong customer/provider relation is built on trust and good history. Changing provider still seems something frightening to the end-user, whose business is only marginally related to Grid (Grid is only one of the supports for his/her business). Placing such an End-User in a competitive marketplace can be daunting, when the user is not mainly interested in peak performance or maximum utilisation, but instead in securing a totally reliable IT infrastructure. In such cases, End-Users can outsource their resource needs to a "Usual Provider" (UP), with whom firm contracts are signed for the long term (this is the scenario of the Business Experiment in Enhanced IMRT planning using Grid services on-demand with SLAs [6]). All the risk is outsourced on the UP. This UP then becomes the traditional user of a Grid schema, as it may face resource scarcity (when its clients suddenly have peak demands). This UP relies on other Grid resource Providers to satisfy the need, but only temporarily, as the UP's resources are sufficient for the usual business. The UP is eager to sign SLAs with other providers for some short periods to cover the peak demands, but is usually self-sufficient. The basic idea is to understand that the provider of a given service can become the client of an outsourced service, and this also applies to SLA client and provider.

References

1. A. Andrieux et al., Web Services Agreement Negotiation Specification (WSAgreementNegotiation) (draft), https://forge.gridforum.org/projects/graap-wg
2. A. Andrieux et al., Web Services Agreement Specification (WS-Agreement), March 2007, http://www.ogf.org/documents/GFD.107.pdf
3. Amazon EC2 Service Level Agreement (Effective Date: 23 October 2008), http://aws.amazon.com/ec2-sla/
4. Amazon S3 Service Level Agreement (Effective Date: 1 October 2007), http://aws.amazon.com/s3-sla/
5. Business Experiment 9 (BE09) Distributed Online Gaming, http://www.beingrid.eu/be9.html
6. Business Experiment 25 (BE25), Business Experiment in Enhanced IMRT planning using Grid services on-demand with SLAs (BEinEIMRT), http://www.beingrid.eu/be25.html
7. I. Foster, Globus Toolkit version 4: software for service-oriented systems, in IFIP International Conference on Network and Parallel Computing. LNCS, vol. 3779 (Springer, Berlin, 2005), pp. 2–13
8. Gridipedia Technical Article on Service Level Agreements, Accessed 29 June 2009, http://www.gridipedia.eu/sla-article.html
9. List of WS-Agreement implementations, from the OGF GRAAP-WG. Visited 15 April 2009, https://forge.gridforum.org/sf/wiki/do/viewPage/projects.graap-wg/wiki/Implementations
10. M.L. Massie, B.N. Chun, D.E. Culler, The ganglia distributed monitoring system: design, implementation, and experience. Parallel Computing **30**(5–6), 817–840 (2004)
11. Nagios Monitoring system, http://www.nagios.org
12. I. Rosenberg, R. Heek, A. Juan, An SLA framework for the GT4 grid middleware, in Collaboration and the Knowledge Economy: Issues, Applications, Case Studies, ed. by P. Cunningham, M. Cunningham (IOS Press, Amsterdam, 2008)
13. Setting Expectations in SaaS, white paper of the Software & Information Industry Association (SIIA), February 2007

14. SLA@SOI, EU FP7 project committed to research, engineer and demonstrate technologies that can embed SLA-aware infrastructures into the service economy, www.sla-at-soi.org
15. S. Taylor, M. Surridge, D. Marvin, Grid resources for industrial applications, in Proc. 2004 IEEE International Conference on Web Services (ICWS 04), San Diego, California (2004), pp. 402–409
16. Web Service Level Agreements (WSLA) Project, http://www.research.ibm.com/wsla/
17. WS-Notification specification, http://www.oasis-open.org/committees/tc_home.php?wg_abbrev=wsn

Chapter 6
License Management

Christian Simmendinger, Yona Raekow,
Ottmar Krämer-Fuhrmann, and Domenic Jenz

Abstract The fact that established license management solutions like FlexNet are not supported in Grid or Cloud environments is one of the main inhibitors for the uptake of Grid or Cloud technology in Industry. In order to resolve the issue the BEinGRID project has initiated a dedicated License Management Technical Cluster. The Cluster has developed design patterns as well as corresponding implementations, which allow the use of currently available technology in Grid or Cloud environments. The solutions are generic and additionally support the transition to pay-per-use scenarios for licensed applications from Independent Software Vendors.

6.1 Introduction

This section highlights some of the results that have been obtained during the course of the BEinGRID project. It begins by defining the Common Technical Requirements. Then we examine some of the most relevant Common Capabilities. These are the attributes that a solution to a Common Technical Requirements must have. Where an implementation is available for a particular capability or to solve a common requirement it is also discussed.

6.2 The Overall Challenge

Small and medium enterprises (SME) from the engineering community stand to profit from pay-per-use HPC Grid scenarios. Very few of these SMEs however maintain their own simulation applications. Instead commercial applications from independent software vendors (ISV) are commonly used with associated client-server based licensing. The licensing software which is the de-facto standard in this area is FlexNet (available from Acresso [1], formerly Macrovision). The authorization of these client-server based license mechanisms relies on an IP-centric scheme—a client within a specific range of IP-addresses is allowed to access the license server. Due to this IP-centric authorization, arbitrary users of shared (Grid/Cloud) resources may access an exposed license server, irrespective of whether or not they are authorized to do so. Secure and authorized access to a local or remote license

C. Simmendinger (✉)
T-Systems SFR, Pfaffenwaldring 38-40, 70569 Stuttgart, Germany

T. Dimitrakos et al. (eds.), *Service Oriented Infrastructures*
and Cloud Service Platforms for the Enterprise,
DOI 10.1007/978-3-642-04086-3_6, © Springer-Verlag Berlin Heidelberg 2010

server in grid environments therefore has not been possible so far. Therefore, the use of commercial ISV applications in grid environments was not possible either.

In order to overcome this obstacle we have designed and implemented a License Management Architecture which supports the entire class of client-server license Management Systems. The solution is generic and thus independent of the different grid middlewares. It is suitable for most grid scenarios and can be employed for local use as well as for clouds.

6.3 Technical Requirements

In this section we describe the technical requirements. We give a brief description of the background and explain innovation and business impact.

6.3.1 Gridification of Currently Used License Management Systems

6.3.1.1 Description

A generic solution was required, which would support all grid middlewares and the entire class of current client-server based License Management Architectures (including FlexNet). The solution also had to be usable in a non-Grid context (e.g. local use, cloud) and hence had to cover a large spectrum of scenarios.

There are some side-implications with respect to accounting associated with the above requirement: whereas in the non-Grid scenario the bill already has been paid in advance and therefore accounting plays a minor role, the pay-per-use model needs to support a flexible cost unit based accounting, rather than an identity bound accounting: The reason is that usually institutions or research groups own the licenses, not their individual members.

6.3.1.2 Innovation

No comparable tool currently exists. The LM architecture bridges the License Management gap and allows commercial ISV codes to be used in grid or cloud environments.

6.3.1.3 Business Impact

The lack of a grid solution for client-server based license management readily implies that the vast majority of users from industry have not been able to use their ISV applications in grid environments. The license management architecture presented here, thus potentially increases the grid market size in the area of engineering on-demand computing by a large factor.

6.3.2 Limited License Service Provider (LSP) Capability

6.3.2.1 Description

Prices of licenses for ISV simulation applications in the area of HPC typically exceed the cost of the corresponding required resources (CPU, memory, filespace) by more than two orders of magnitude—a single license can cost up to 100,000 Euros per year. In contrast, a CPU hour sells for as low as 10 cent. Therefore a pay-per-use model for licenses is required in order to provide a satisfactory on-demand computing scenario with a licensed application from an ISV.

Licenses for ISV codes are typically issued on a yearly basis: Customers buy a fixed number of licenses, with associated features and included support. Therefore the generated revenue for the ISV is predictable and stable. In addition, this business model guarantees that the provided codes are always in line with the requirements of the end-users: there is a close dialog between ISV and end-user.

Unfortunately, the business model is in contrast to a pay-per-use scenario. In a pay-per-use scenario there is no predictable revenue for the ISV and unless the ISV is also the license service provider (LSP), the ISV would lose the direct contact to their end-users.

The contrasting business models make a direct transition towards a pay-per-use license model on a new technology basis rather unrealistic. Instead, ISVs will need to constantly evaluate and refine the evolving new business models in a non-interruptive transition on the basis of currently used technology.

In this transitional phase typically the ISV will assume the role of the LSP. Alternatively, the ISV might decide to outsource this role to a service provider. This scenario is what we refer to as limited LSP capability and any pay-per-use License Management solution would have to support this transition.

6.3.2.2 Business Impact

The currently established business model implies a substantial over-provisioning of licenses: End-users need to buy more licenses than they require on a daily basis, in order to satisfy their peak requirements. With a pay per-use scenario this over-provisioning immediately becomes obsolete with a corresponding loss of revenue for the ISVs. Therefore nearly all attempts to convince ISVs to agree to provide pay per use business models have been unsuccessful in the past.

On the other hand, a pay-per-use model would create a new source of revenue for ISVs, because SMEs, which so far could not afford to purchase licenses would become able to access the licenses on a pay-per-use basis. Additionally, large customers become able to dramatically increase the number of licenses during peak-demand periods.

6.4 Common Capabilities

Industrial environments typically rely on commercial applications of ISVs with an associated License Management—usually FlexNet from Acresso [1], which is the de-facto standard in this area. FlexNet has a closed API, is proprietary and based on a simple client-server mechanism. The FlexNet scheme allows floating licenses, which are not bound to a specific host. Rather they are allocated dynamically to arbitrary hosts. Licenses are checked out at the license server when an application starts and checked in when it ends. Therefore FlexNet is suitable in principle for usage in a Grid environment. There are, however, major security and identity issues with respect to the access to the license server in a Grid environment. For example the FlexNet software is able to filter legal and illegal accesses based on the host IP, but is not able to grant access on the basis of user/group certificates. This implies that on every Grid site an unauthorized user could check out and use an arbitrary number of licenses once the corresponding license server is exposed. In order to support this standard (or in general: any client-server license management scheme) we propose to transparently reroute the encrypted socket-based communication between client and server via a SOCKS [7] proxy-chain. The communication from the license client then can be transparently forwarded (via a SOCKSified job shell) to a remote upstream proxy and then to the remote license server. The run-time authorization at the upstream proxy is handled via a PIN and associated encrypted one-time passwords (TANs). The PIN here represents a license account and can be used to provide the accounting context. License owners (typically institutions) can set up an arbitrary number of these license accounts under a billing account. This mechanism allows institutions or research groups to share access to licenses and to use licenses in a cost unit based accounting context. A self-imposed budget-control for pay-per-use scenarios is possible. The handling of the one-time passwords (generation of TAN lists, license accounts and their properties) is implemented as a Web service. Additionally, a portal may be used, which enables users to access these Web services, conveniently share accounts, automatically extract one-time passwords, submit correspondingly modified jobs, and provide accounting facilities.

The entire system consists of four entities: the provider of compute resources, an SME, an engineer at the SME and a license provider. The SME can have various license accounts that correspond to internal projects or departments, facilitating their internal accounting and monitoring of expenses. Hence, these license accounts are not managed directly by the engineer at the SME, instead the engineer is provided with the account and a certain number of one time passwords that he/she has to his/her disposal for submitting jobs. Simultaneously an engineer can run an identical job under different license accounts, where the individual license accounts usually would be assigned to a cost-unit.

License Management Architecture

Most of the capabilities in the LM architecture can be assigned to specific components. Figure 6.1 shows the entire LM architecture.

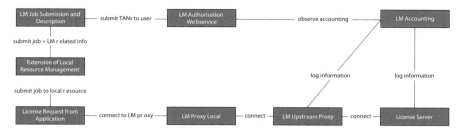

Fig. 6.1 High level view of LM architecture

6.4.1 LM Authorization

6.4.1.1 Purpose

This capability covers authorization mechanisms required with respect to license resource requests, namely whether or not a user is entitled to use a specific license server, specific features of the licensed software or whether limits exist with respect to the number of licenses. In close analogy to access to other resources, possible solutions range from simple locally maintained access control lists or PIN/TAN mechanisms up to a full integration into identity management systems like Shibboleth or VOMS with explicit requests to home organizations of the users and/or third party license service providers or even requests to license Brokers.

6.4.1.2 Architecture

The LM authorization has interfaces to the "Job Submission and Description" capability. It also interfaces with the encapsulation of the License Server. The latter capability encapsulates the server and grants the actual access to the license server. In the implemented LM architecture this latter capability is provided by the LM Proxy Chain and then LM Accounting.

Functionality

In order to provide support for currently existing client-server based License Management systems, authorization is required when a submitted job runs. A further challenge is that in most grid middlewares, certificates are not available at job run time. A certificate based authentication and authorization therefore does not seem suitable.

Moreover the authorization needs to be generic with support for all middlewares. It needs to be suitable for local use, grids and clouds. Finally, authorization has to be suitable for hostile environments in the sense that a single lost credential must not compromise the security of the entire system.

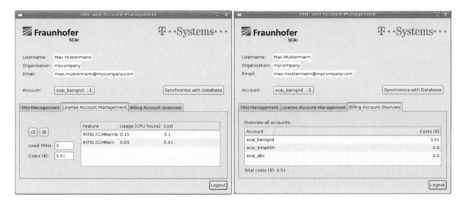

Fig. 6.2 LM architecture—GUI

6.4.1.3 Implementation

The LM implementation includes two Web services. One is responsible for authorisation and handles the one time passwords. The Web service is created using Axis2 and deployed at the site of the provider. The corresponding clients run on the site of the SME.

The client consists of a graphical user interface that allows convenient management for all license accounts that a SME holds for a given system provider site. Alternatively the client may be embedded in a JSR168 compliant portal.

TAN Web Service

The Web service handles the one time passwords that are submitted together with the job description. It provides functionalities to create new TAN lists, to block and unblock TAN lists and to check how many TANs have been used from the current list.

The core element here is the tool pam_sotp [4] which is a UNIX tool that allows the creation of TAN lists. This tool is encapsulated by the Web service. Features of the tool that are accessible via the Web service are the following:

- Create TAN list: This function creates a TAN list, i.e. a list of one time passwords. The user can specify the required number of TANs. The number of TANs corresponds to the number of licenses the user wants to request later on. The pam_sotp tool allows to specify over which alphabet the TANs should be created and which length they should have. These parameters are hardcoded in the Web service and are not the choice of the user. An easy extension of the component could be provided, if required.
- Block TAN list: If a license account is exceeding a budget constraint (e.g. monthly budget) the TAN list for this given license account can be blocked. The TAN list remains unchanged, but all attempts to access it result in an error.

- The user can unblock a previously blocked TAN list. This allows the license account that corresponds to this TAN list to access this TAN list again. The main purpose here is to allow for an automatic self-imposed budget control.

6.4.1.4 Validation Scenario

Grid Friendly License Management

The organisation BSYS owns a certain number of licenses for an ISV Code, runs a corresponding FlexNet License Server and wants to use these licenses for calculations at a grid resource provider. Since BSYS uses the ISV Code for a large number of projects, BSYS requires a cost-unit accounting context within which the calculations are performed. The resource provider might not be known at job submission time. In order to perform the calculation with the ISV Code, organization BSYS now would have to open its firewall and allow any potential remote grid site to access its license server. The LM architecture resolves this problem by authorizing access to the license server at BSYS via a PIN/TAN mechanism.

The organization BSYS also wants to access additional pay-per-use licenses for peak demands. To that end they sign a frame-contract with the ISV. Similarly to the above case the ISV requires authorized access from shared grid/cloud resources to these pay-per-use licenses.

For a pay-per-use scenario an ISV also needs a fine grained accounting and billing, which is able to resolve details about requested licenses (wallclock, features) and also is able to provide the accounting context within which these licenses and features were used.

Once the frame-contract has been signed a billing account is set up. The billing account includes all required details about organization, email etc.

This billing account grants the organization BSYS the right to set up an arbitrary number of license accounts. The access rights of these license accounts are inherited from the billing account, but can be pruned upon request. Typically a license account would correspond to a cost unit. The billing account can request a TAN list (for each license account), which can be split among all engineers who need to run (and account) their simulations under that cost-unit.

The engineer provides the correct accounting context through an environment variable and submits the job. At run time a locally stored part of the corresponding TAN list is accessed, once the application tries to connect to a remote license server. Both types of licenses (owned by BSYS and pay-per-use from ISV) can be used—simultaneously, if required.

After successful authorization the accounting process is triggered and at the end of the session both elapsed wallclock time and details about used features are logged to a database.

At any time, organization BSYS can access the LM Web services for accounting and billing and retrieve an accumulated view of license hours and price—per license feature and/or license account.

At pre-defined intervals the ISV then retrieves the accounting information from the database and issues a corresponding bill. The assigned cost can vary—depending on the type of the license account (e.g. license used in academic or commercial context).

Finally the organization BSYS assigns both the external bill from the ISV and the accumulated accounting information from the internal license server to the proper cost-unit.

6.4.2 Extension of Job Description and Submission

6.4.2.1 Purpose

This capability covers the extension of the job description and its submission with respect to license management. A user needs to provide details about the requested licenses, including authorization as well as the accounting context. This capability allows a user to request license resources in a similar manner as currently implemented in Grid middlewares for any other resource (CPU, memory, etc.). The resources here can be either own licenses, licenses provided by the service provider or an external LSP. The resources can be requested and used simultaneously. In many client-server License Management systems the application polls for a license in a pre-defined sequence until the request can be satisfied. For example this allows companies to use additional pay-per-use licenses, only if their own licenses are currently not available. In order to make cost-efficient use of licenses, information about the use of licenses need to travel with all corresponding jobs.

This allows e.g. to use additional pay-per-use licenses, if own licenses are currently not available, or to co-schedule licenses and resources. The capability so far did not exist.

6.4.2.2 Architecture

The extension of job description and submission extends the interface to pass details about authorization, accounting context and license resources to the application. The extension of Job Submission possibly needs to interface with a corresponding extension for the Resource Management System and the LM Authorization, if e.g. Authorization is based on local Access Control Lists (ACL) or licenses are bound to specific nodes (node-locked licenses).

6.4.2.3 Implementation

The job description for job submission has been extended: Within the job description the user has to provide his/her license account name—which simultaneously provides the accounting context—via the LICENSE_ACCOUNT environment variable.

Additionally, the job input needs to include one or several TANs—e.g. as a part of the corresponding global license account TAN list. This partial TAN list also can be stored locally. If available, the correct TAN list is automatically chosen through the LICENSE_ACCOUNT environment variable.

6.4.2.4 Validation Scenario

See Sect. 6.4.1.4 Grid friendly license Management.

6.4.3 Resource Management Extension

6.4.3.1 Purpose

This capability covers the extension of a local resource management system. This is not mandatory and only required if there is no additional run-time authorization at the license server.

In a batch prologue the resource management can dynamically reconfigure the access from the assigned resources to an external network resource. At this point security of access to the license server is transferred from a certificate based authentication/authorization level to the required network level security at which e.g. FlexNet operates.

6.4.3.2 Architecture

The Resource Management Extension interfaces with the Job Submission and Description and a local access control list or database. Additionally access from locally assigned resources to external resources has to be reconfigured dynamically on a per-job basis.

6.4.3.3 Implementation

A local SOCKS proxy (at the service provider site) is re-configured on a per-job basis by the local resource management systems, thus providing an additional layer of security. If only local authorization is used, Information from FlexNet itself needs to be synchronized with information from the authorization module in order to provide a complete cost unit based accounting.

6.4.3.4 Validation Scenario

See Sect. 6.4.1.4 Grid friendly license Management.

6.4.4 Encapsulation of License Server

6.4.4.1 Purpose

This capability relates to the integration of existing license servers. It not only addresses a possible encapsulation as a Web service but more generally, the integration of the license server into the respective Grid middlewares. We note that in our license management architecture implementation the LM proxy-chain, together with the LM monitor, provides this encapsulation.

6.4.4.2 Architecture

The encapsulation of the license server requires interfaces to the LM Job Submission. It also needs to support both LM Authorization and LM Accounting. The details of this encapsulation strongly depend on the underlying License Management System and the chosen transport layer. For the FlexNet License Management and the SOCKS transport layer, details are provided in the implementation section.

6.4.4.3 Implementation

The core element of the Encapsulation of the License Server is a standard SOCKS 5 proxy (Circuit Level Gateway). The user provides a license account name with corresponding one time passwords (TANs). These TANs are validated when the upstream proxy is accessed. To that end the initial connect() from the application is replaced with a connect() over the proxy chain (via the LD_PRELOAD variable). For our purposes we identified the combination ss5 and tsocks [6] to be the most suitable, since it allows a chaining of proxies. Once the job runs at the provider site and the license is required, the connection via the proxy chain to the license provider is automatically established. The license request is encapsulated on the SOCKS network layer and both the LICENSE_ACCOUNT and a TAN are attached as username and password. The TAN is evaluated at the remote license server. If access for this specific license request (IP:PORT) can be granted for the given license account the request is forwarded to its destination.

Additionally, before forwarding the request, the accounting process is triggered after successful authorization.

6.4.4.4 Validation Scenario

See Sect. 6.4.1.4 Grid friendly license Management.

6.4.5 Accounting and Billing

6.4.5.1 Purpose

This capability covers the accounting and billing of licenses. In order to produce the complete accounting information, log information from both the proxy (accounting context/time-stamp) and the license server (time-stamp, number of licenses, license features) are required. The actual details of billing and accounting might depend on the underlying business model. Depending on whether licenses are owned by the user, the service provider, an external static LSP or obtained via a Grid broker, the exchange and assembly of the actual accounting and billing information will differ.

6.4.5.2 Architecture

Accounting and billing is triggered by the encapsulation of the license server. It also has an external user interface. In order to support e.g. self-imposed budget control, this interface can serve to automatically lock the license account, if the assigned budget (per week or month) is exceeded.

6.4.5.3 Implementation

A Web service is responsible for accounting. This Web service provides functionalities for cost overview. The actual accounting is triggered by the LM Proxy chain. To that end the requests to the license server are serialized at the proxy and an initial query to the licenses server is performed. After a pre-defined negotiation time between license client and server a second query is performed. The difference between initial and second query determines number and type of requested features. When the connection finishes wallclock and requested features are written to a database.

Accounting Web Service

This Web service allows the user to check the license usage of all accounts held. The functionalities of the accounting Web service are as follows:

- Get license accounts, this functionality retrieves all license accounts that a user manages from the server.
- Create license account: allows a user to create a new license account.
- Delete license account: deletes an existing license account.
- Get account info of a selected license account. This information consists of the number and walltime of license features used so far. It can also provide information on how much money a license account already spent for licenses. Furthermore a user can see how many TANs have been used and how many are still available.
- Additionally, a view over all license accounts is provided.

6.4.5.4 Validation Scenario

Application Service Provisioning

Organisation CSYS wants to run a Job with an ISV Code on the resources of a service provider. They need to acquire the corresponding licenses from this Service Provider (ASP model). They register a billing account at the ASP, set up an arbitrary number of License Accounts under that billing account and retrieve the corresponding credentials (License Account/TAN).

Before the job is started, the resource management system queries the LM monitor whether there are sufficient licenses available for this job. If the query succeeds the job starts. The ISV application then tries to access the license server at the site of ASP. The request is re-routed over the proxy chain to the license server. In the first stage (re-routing to the local proxy) the currently assigned License Account is extracted from an environment variable of the job and attached to the license request (SOCKSification). The request is forwarded to the upstream proxy. The upstream proxy authorizes the request and writes an accounting record based on information it retrieves from the license server. The request is forwarded to the license server where it is granted and returned—via the proxy chain—to the application. At job end the accounting record is finalized and assigned to the license account. Then the record is assigned to the billing account and a cost-unit based accounting and billing is provided.

6.4.6 LM Monitor

6.4.6.1 Purpose

This capability covers the ability to co-schedule licenses and resources in the following way: jobs are not started before a required license is available. We remark that in most cases this can not be a true co-scheduling since the underlying client-server License Management would need to support this feature. Instead the local resource scheduler polls for licenses until the license requirements can be met by the license server. The component only interfaces with the local scheduler and an external Web service which in turn queries the license server.

6.4.6.2 Architecture

The component only interfaces with the local scheduler and an external Web service which in turn queries the license server. Apart from a query from local schedulers, information from LM monitor also can be used as input to a license broker.

6.4.6.3 Implementation

The LM monitor is implemented as an independent (from LM Architecture) Web service.

6.4.6.4 Validation Scenario

See Sect. 6.4.5.4 Application Service Provisioning.

6.5 Conclusion

In this chapter we briefly sum up the key elements from lessons learnt and give a few recommendations—both for the use of the LM components and License Management in distributed environments in general.

6.5.1 Lessons Learnt

6.5.1.1 Support for License Management in Grid Environments

Prices for licenses typically exceed prices for CPU resources by two orders of magnitude. To give an example: A single license for a structural mechanics code can cost up to 100,000 Euro per year. In comparison, the corresponding CPU resource can be obtained for less than 1,000 Euro a year.

All grid solutions so far have addressed this latter topic: Optimization of usage of standard resources like CPU and filespace. None has addressed the former. Since most application used in industry—especially in the area of engineering and HPC—are commercial ISV codes, the lack of support for License Management in grids can be seen as one of the main inhibitors for the industrial adoption of grid technology [2, 3].

6.5.1.2 Transition to Pay-Per-Use

The LM architecture provides a solid technological basis for a transition towards a grid-friendly pay-per use model. However, in order to achieve this long-term goal, the ISVs need to re-think their pricing strategy, investigate pay-per-use scenarios and evaluate corresponding business models.

In the long run—after a transitional period towards pay-per-use—the commercial exploitation of Grid will need a different technology: FlexNet has severe limitations, even if used in the context of the LM Architecture presented here; most notably it lacks support for a meta-scheduler. Once a satisfactory—both for ISVs and end-users—pay-per-use model has been established, a better technology will be required. To this end the BEinGRID license management technical area has initiated several meetings with the FP7 STREP SmartLM [5].

6.5.2 Recommendations

6.5.2.1 Transition to Pay-Per-Use—the Business Model Aspect

Since ISVs potentially might lose revenue—and maybe even more problematic: predictability for revenue—in a pay-per-use scenario, the transition to pay-per-use will take a long time and ISVs will be very reluctant to support this scenario. Therefore it is very likely that end-users need to push the ISVs towards that direction—either by resorting to open source solutions or by insisting on this type of business model in price negotiations.

General recommendation: Grid license management is not required, but rather grid friendly license management. A successful license management architecture will need to be generic. It has to be grid-friendly and at the same time it has to work in a non-grid context. The end-users will want to use their licenses in very different environments—on a PC, on a local cluster, in a grid scenario—and the license service needs to support all these environments. Additionally, the license providers will very likely support only one such service—and this service has to serve all the different purposes.

6.5.2.2 Recommendations for Use of LM Components

The License Management Architecture can simultaneously support a variety of scenarios. A single license SOCKS/Web Service simultaneously can provide licenses for local use on a PC, it can serve intra-grids (e.g. Local Use scenario: Linux clusters where CPU resources are shared between license owners and other users) but it also can provide secure access to the license server from a remote Grid site, irrespective of the underlying grid middleware. The main differentiating factor between these scenarios is the respective authorization and the provided accounting context. For grid middlewares or portals both of these aspects are closely coupled to the job submission and description.

For general use, we would recommend to use the entire LM architecture, since it has the additional benefit of delivering a cost-unit based accounting. For scenarios where such a feature is not required and/or explicit trust delegation is possible, we suggest the use of LM Proxy Standalone component—possibly with the modification of access control lists in a database instead of a the currently used flat file hierarchy.

References

1. Acresso Software, http://www.acresso.com
2. Y. Raekow, C. Simmendinger, O. Krämer-Fuhrmann, License management in grid and high performance computing, in International Supercomputing Conference (2009)

3. C. Simmendinger, Y. Raekow, O. Krämer-Fuhrmann, H. Herenger, Support for client-server based license management schemes in the grid, in Collaboration and the Knowledge Economy: Issues, Applications, Case Studies, ed. by P. Cunningham, M. Cunningham (IOS Press, Amsterdam, 2008)
4. Simple One Time Passwords (OTP) authentication for PAM (PAM_SOTP), http://www.cavecanen.org/cs/projects/pam_sotp/
5. SmartLM European FP7 project—Grid-friendly software licensing for location independent application, http://www.smartlm.eu/
6. Sockets Server SS5, http://ss5.sourceforge.net
7. SOCKS Protocol Version 5, RFC1928, http://tools.ietf.org/html/rfc1928

Chapter 7
Data Management

**Craig Thomson, Kostas Kavoussanakis, Mark Sawyer, George Beckett,
Michal Piotrowski, Mark Parsons, and Arthur Trew**

Abstract Data management is an important area of Grid research. It is concerned
with the storage, access, translation and integration of data. The Data Management
Technical Area of the BEinGRID project was set up to analyse the data requirements
of pilot projects and to support them in using data management related middleware.
After identifying these requirements it also developed design patterns to provide a
guide to other businesses which may face similar problems in the future. As a further
aid to businesses interested in adopting the Grid, the technical area also extended
existing middleware to allow it to implement some of the identified design patterns.

7.1 Introduction

Data management is an important area of Grid research. It is concerned with the
storage, access, translation and integration of data. It hopes to answer questions
like:

- Where should I put my data?
- How should I get to it?
- How do I present my data in a way others will understand?
- How can I combine data from different places?

All of these questions are important to modern businesses. In many industries,
collaboration and the efficient flow of information between organisations is critical.
For example, just-in-time techniques [14] aim to improve the efficiency of a supply
chain and to do this effectively they need access to up to date information from
multiple organisations.

One of the most familiar definitions of "The Grid" is a Computational Grid [15]
in which multiple distributed computer systems calculate a common result. One of
the things which characterises Grid computing is the heterogeneity of the computing
resources used. These differences can be in the hardware, the software or both and
interaction between the different resources is required for a Grid to be usable. A similar definition can be applied to data and Data Grids; data from multiple sources can

C. Thomson (✉)
EPCC, The University of Edinburgh, James Clerk Maxwell Building, Mayfield Road, Edinburgh
EH9 3JZ, UK
e-mail: c.thomson@epcc.ed.ac.uk

T. Dimitrakos et al. (eds.), *Service Oriented Infrastructures
and Cloud Service Platforms for the Enterprise*,
DOI 10.1007/978-3-642-04086-3_7, © Springer-Verlag Berlin Heidelberg 2010

be combined or used to produce a desired result or effect. In the case of data, the heterogeneity can extend to the storage format as well as the machine type and the software. Considerations such as the structure of data or the query language (xml, SQL, files) as well as the particular software product (Oracle, MySQL, etc.) are important additional considerations for a Data Grid and make this a very rich and complex problem area.

The Data Management Technical Area of the BEinGRID project was set up to analyse the data requirements of a number of pilot projects termed Business Experiments (BEs) and to support them in using data management related middleware. In addition it analysed the problems the BEs faced and extracted the common requirements multiple BEs had. After identifying these problems and requirements it also developed design patterns to provide a guide to other businesses which may face similar problems in the future. As a further aid to businesses interested in adopting the Grid, the Technical Area also extended existing middleware to allow it to implement some of the identified design patterns. The focus of these middleware modifications was OGSA-DAI [10], a long-standing data access and integration middleware currently developed at The University of Edinburgh as part of the OMII-UK project [11].

This chapter highlights some of the results that have been obtained during the course of the project. It begins by defining the common technical requirements. Then we examine some of the most relevant common capabilities. These are the attributes that a solution to one of the common technical requirements must have. Within the discussion of common capabilities we will also describe particular design patterns which can be applied to solve these problems effectively. Where an implementation is available for a particular capability or to solve a common requirement, it will also be discussed.

The results produced by the Data Management Technical Area came from the analysis of concrete BEs in business sectors. Not all of the experiments had a strong interest in data management and therefore the conclusions we have drawn are based on our experiences with a subset of the experiments.

7.2 The Overall Challenge

The prominence of data in the majority of the BEs shows how important information is in a modern business environment. The sheer variety of uses of data was a challenge in itself when it came to analysing the different requirements of the many BEs. It is difficult to give one single view of the challenge for data management. Instead we list below some of the important areas which are relevant to the business areas the project has investigated. These are as follows:

- Data transfer
- Integration of data from different organisations
- Replication of data between organisations
- Heterogeneous data.

The focus of the work of the Data Management Technical Area has been to address the challenge of accessing and utilising existing data. Several BEs had existing sources of information which they were using in a manual way. BE12 (Sales Management System [1]) aimed to make existing stock, pricing and sales information at franchises available at the head office. BE24 (Grid technologies for affordable data synchronisation and SME integration within B2B networks—also called GRID2(B2B) [2]) dealt with existing B2B networks which it wished to enhance with automated data exchange.

It is usually much easier to develop a new computer system rather than integrate a number of existing systems which may not naturally fit together. This extends to dealing with data, as the requirements and areas of interest for the BEs showed. The problem of dealing with existing data sources is a real challenge for applying Grid data management middleware to business scenarios.

Driven by the requirements of the BEs, three software components were developed: the Data Source Publisher; the Query Translator; and the OGSA-DAI Trigger.

By developing the Data Source Publisher, we have tried to make the data integration and access capabilities of OGSA-DAI more easily available. This will help to reduce the differences between sources of data as it provides a layer of abstraction. The Query Translator (see Sect. 7.4.2.2) works in situations with multiple sources of data in different formats and it makes it easier to access and manipulate them. The OGSA-DAI Trigger component allows new processing based on OGSA-DAI workflows to be integrated with databases populated by other applications.

The key benefit of these components is to provide solutions which move away from particular database implementations and allow for a more generic and flexible approach to data which is more concerned with extracting and using the data rather than being constrained by its particular format.

7.3 Technical Requirements

The first part of our analysis of the BEs focused on identifying the common problems faced. The BEs were selected from a number of different industries and were trying to solve different, specific, business problems. Our aim was to identify the common points that existed in this diverse set. By identifying these Common Technical Requirements we gain an understanding of the problems of interest to businesses. It also allows us to focus on identifying solutions to the most important questions businesses trying to adopt Grid techniques for Data Management are facing.

There are a number of common requirements which were featured in the use case analysis of the BEs related to data management. In addition the experience of technical activities of the BEinGRID project has highlighted some very important requirements which need to be addressed to expand the use of Grid and Grid Middleware by business. They are as follows:

- Accessing data from different locations

- Accessing heterogeneous data
- Respond to changes of data in a database.

In addition it was clear that secure transfer of data is vital in a business environment. But this problem is broad enough to warrant its own Technical Area, so it will not be discussed further here.

These requirements are clearly very broad in scope and have been addressed to varying degrees by existing middleware. As part of the analysis process we discovered that OGSA-DAI was the most frequently used middleware for data access in the experiments so it became the focus of the component development. Our aim has been to enhance the OGSA-DAI middleware to enable it to meet the requirements we identified from the BEs.

7.3.1 Accessing Data from Different Locations

A fundamental requirement for many Grid applications is the need to access data from more than one location. When examining this requirement there are two important facets, as follows:

- Remote access to data
- Accessing more than one source of data.

Both of these points are important as they help us to understand the real benefits that come from addressing this requirement. Remote access to data allows businesses to manage information centrally. For example they can generate sales figures easily or coordinate pricing for all the outlets of a company from the central office. Accessing more than one source of data allows for greater access to information, allowing more complex algorithms to be used for scheduling or analysis. It also allows multiple companies to pool information to improve the results of their analysis.

When examining access to data it is interesting to note the type and extent of access required. Data exists somewhere on a computer system. Remote access can refer to accessing data outside the application it exists in, for example finding a new use for data which is held inside a piece of stock-control software. It can also refer to using the data on a different machine in the same organisation. At the widest level it can refer to using the information of another organisation. The level of access required has implications on the level of security needed to protect the data and on the amount of control the users will likely have over the data.

7.3.1.1 Business Benefit

The business benefits for enabling access of data in different locations are in part related to the business opportunity. However, improving access to data has the following general benefits:

- New opportunities for collaboration with different organisations may be found.

- Better results may be generated from access to more sources of information.
- Costs may be reduced due to better integration of data across multiple sites.
- Larger markets for products and services may be accessed.

7.3.1.2 State of the Art and Innovation

There are a number of existing technologies which allow access to data in different locations (Web Services, FTP, OGSA-DAI [10], sockets, JDBC [8]). However there are still open questions. In many situations where data is owned by an external party the users can have little control over the mechanism used to make the data available. In this case there is scope for providing a layer that translates between different methods of exposing data.

Part of the problem when dealing with distributed resources is the additional complexity they add to any solution. The application has to deal with data in multiple sources, which may change over time; the number of resources may also change over time. A possible area of innovation is providing a layer which abstracts this complexity away from the application. By adding further layers of abstraction the higher level business logic can be made simpler and the more generally useful capabilities can be reused in multiple applications.

7.3.2 Accessing Heterogeneous Data

As soon as data is accessed from multiple locations, there is a likelihood that the data will be in different formats. There are many ways to store data: files, xml databases, relational databases. Even with support for SQL, different vendors interpret the standards differently and there are many versions of the standards. On top of those problems there is also the question of the data format itself. The tables in a database may be arranged differently. There are a variety of different file formats or the data could be held in an entirely proprietary format. The underlying mechanism for accessing data can also be different, not all data is held in files; some could come directly from sensors for example.

The challenge for Data Grids is to provide some mechanism to help reduce the level of heterogeneity to make the data easier to use.

7.3.2.1 Business Benefit

By solving the problems of heterogeneous data we can achieve a number of benefits:

- Development costs may be reduced because applications can be simpler.
- Larger markets can be accessed as data related products can be targeted at more types of data.
- New opportunities for collaboration with different organisations can be realised.
- Better results may be generated from access to more sources of information.

7.3.2.2 State of the Art and Innovation

There are a number of different solutions available which can help handle heterogeneous and distributed data sources. SRB [12] and its successor iRODS [7] are
aimed at generating a robust distributed file system based data store that can handle
heterogeneity at the level of the hardware. OGSA-DAI [10] provides a workflow
language for performing data access and transformation which can help hide many
of the differences between data sources from the end-users. SwisSQL [13] provides
translation between different dialects of SQL. Many of the large database vendors
have products and consultancy aimed at helping organisations to migrate their data.

One solution to some of the problems of heterogeneous data is to provide data
warehousing, where an intermediate data store is used. All input data is transferred
to the warehouse, and transformed into one common format to make the data easier
to query.

An alternative is data federation which keeps the data in its original location and
instead provides an interface layer to hide the distributed nature of the data. It is
also possible to provide different translations in this layer for different data sources.
DiGS [3] is a distributed-data management system that allows one to combine third-
party (that is, in different administrative domains), commodity storage resources—
such as RAID systems and Storage Area Networks—into a large-scale, unified file
repository. DiGS uses lightweight software modules, called Storage Element Adaptors, to hide the complexities of heterogeneity and geographic dispersion from the
client application.

7.3.3 Respond to Changes of Data in a Database

Businesses which are interested in adopting new technology for data management
will already have applications or sources of data. An important requirement is that
they can use these existing data sources and methods of data entry. In order to build
on top of existing infrastructure, a mechanism to respond to changes in a database
would be very valuable. For example, a travel agent and a tour operator might be
partners but each might have a different customer database. If a customer books a
tour at a travel agent, their information needs to be entered twice, once for each
system. It would be better if relevant customer data was automatically transferred
when it was inserted in either database.

7.3.3.1 Business Benefit

Again a solution to this requirement has benefits which are dependent on the business opportunity. It is certainly the case that there is the potential for benefits such
as the following:

- Development costs may be reduced as existing applications can continue to use
  the database as before.

- New markets may be accessed as software can be developed to work alongside existing software or to extend competitors' applications.

7.3.3.2 State of the Art and Innovation

There are a number of replication solutions such as the High Volume Replicator [5]. They provide scalable replication which will transfer information from one database to another. This provides the solution for some applications and is ideally suited to disaster recovery and off-site backup. Many databases also offer replication solutions which allow data to be automatically propagated from one database to another.

Many implementations of SQL provide triggers which allow some actions to be taken in response to changes in tables in the database. They are typically restricted to running SQL commands though it is possible to achieve more with user-defined functions.

Having a consistent mechanism which would allow actions to be taken in response to changes within databases would be a valuable addition to the capabilities of Grid middleware. Providing a mechanism at the middleware level also reduces the level of access required to the database, an important consideration in many cross-enterprise business scenarios.

7.4 Common Capabilities

After identifying these important requirements for business users of Grid it is instructive to consider possible solutions independent of particular software. This helps to make the concepts clear before investigating the implementation details.

There are a number of common capabilities on which it is useful to focus when considering the previously mentioned common technical requirements. These capabilities are as follows:

- Access remote data sources
- Homogenise data sources
- Synchronise multiple data sources.

7.4.1 Access to Remote Data Sources

Often in business, information is held in a different location to where processing will occur. A useful capability is one which allows remote partners to access this data. This is a possible solution to the "Accessing data from different locations" requirement.

This common capability works by passing a message or series of messages between the central and remote systems. The remote system is able to accept a query

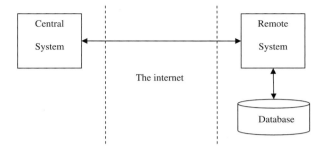

Fig. 7.1 Accessing a remote data source

about its local database, access the database to retrieve that query and then pass the results back to the central system.

A key aspect is that the remote system acts as a translator between the messages that go between it and the central system, and those that go to and from the database. The database does not need to know about the Internet, it is hidden from that by the rest of the system.

7.4.1.1 Purpose

It is often useful for business to be able to access data held in a remote location. This could be because they are collaborating with another business, or because the business has merged or changed.

This capability allows the existing data source to become available over the Internet without having to change it. The capability is useful for all businesses requiring access to a remote data source and helps encapsulate legacy systems that are not Grid-enabled.

Its business value stems from Grid-enabling legacy systems.

7.4.1.2 Implementation

Existing Solutions

Depending on the particular scenario, there are many different ways of making data available remotely. Low level software using sockets or web services can be written. It is also possible to use existing applications, middleware or higher level libraries to provide more features. GridFTP, FTP, JDBC or ODBC are all possible methods of exposing data, as is middleware like OGSA-DAI.

The OGSA-DAI project aims to provide the e-Science community with a middleware solution to provide access to and integration of data for applications working across administrative domains. OGSA-DAI provides access to a variety of different database types and allows data to be published via a web service interface. It also contains a variety of activities which allow data access, transformation and delivery.

Data Source Publisher Design Pattern

To provide a more detailed view of potential solutions to a particular technical requirement a number of design patterns have been developed.

The Data Source Publisher pattern describes a mechanism for allowing data to be made available for access at another location. It also provides a layer which can be used to translate or abstract the data type. The goal is to allow an existing system to be Grid-enabled so that it can be accessed via other Grid middleware components.

The pattern works by adding a component which communicates with the existing data source. This component provides another interface which allows the information to be accessed remotely. The intention is that this pattern allows any existing applications to use their existing procedures to access data.

More information on this design pattern can be found on Gridipedia [4].

Component Development

As part of the work of BEinGRID a number of components have been developed to address the data management requirements. One of these components is the Data Source Publisher which implements the "Access to Remote Data Sources" capability using the OGSA-DAI middleware. The goal of the Data Source Publisher is to simplify and automate the deployment procedure of OGSA-DAI.

The Data Source Publisher provides a simple, GUI-based installer which deploys OGSA-DAI and publishes a data source via web services. It is much more convenient and requires much less effort on the part of the person installing the middleware if everything they need is bundled together and can be installed in a few simple steps. Installation and deployment of OGSA-DAI on a computer requires the download and installation of correct versions of its pre-requisites as well as database drivers.

By using the Data Source Publisher these requirements are reduced and the process to install OGSA-DAI is simplified, since everything is handled inside the GUI installer. There is a single download which contains the correct versions of all the software required to set up OGSA-DAI. All that is required is to configure the component for the application.

More information on the Data Source Publisher can be found on Gridipedia [4].

Outstanding Issues

It is still difficult to answer the question of what is the appropriate technology to use to expose existing data. That depends both on technical and business requirements and organisational constraints and knowledge. What is certain is that in order to be considered as a viable solution, Grid technology must be accessible. One of the first barriers to adopting a piece of technology is ease of installation. If software is made of many pieces and takes many steps to install, a failure of documentation or user error at any of these steps can get the solution discarded at the evaluation stage. This is a problem for all middleware, not just data management related software, but it is vital that the software developed in this and other projects be accessible to a novice user. This is part of the motivation behind the development of the Data Source Publisher but much work still remains to provide straightforward set up and configuration procedures for Grid software.

7.4.1.3 Example of Use

BE12 [1] wanted to make data from heterogeneous databases on pizza franchises available to the head office. They elected to address this problem using OGSA-DAI. At the start of the experiment there was no way to access the data outside the franchise. The mechanism to deploy any upgrades to the franchises involved the engineers who installed the existing tills and computing infrastructure. They were not Grid experts and the goal was to make the set up process as intuitive and simple as possible for them. In particular the process needed to be easily repeatable at the many outlets.

To address this problem a custom installer was produced which deployed the precise version of OGSA-DAI required and their software. This installer formed the basis of the Data Source Publisher developed as part of the project.

7.4.2 Homogenise Data Sources

This capability relates to the remote access common capability and shares some of the same principles. The goal is to present data with different structure in the same way. There are two primary ways to do this:

- Copy the data and transform it into a common format.
- Add a layer on top of the data sources to hide the differences between the data.

The first involves collecting data into a central database and converting it all to a common format. In BEinGRID, there has not been a strong requirement for this form of data manipulation.

The second option has emerged as being more interesting to the BEs. The aim here is to overcome the problems of dealing with a variety of data sources such as databases and files. By using a component which abstracts the data access into an interface that is compatible with all of the different data sources, these differences can be hidden from an end user or application.

Note that despite their differences both options rely upon the provision of translation components to handle the conversion to the common format and also a common interface to support the homogenised access to the resources for the client.

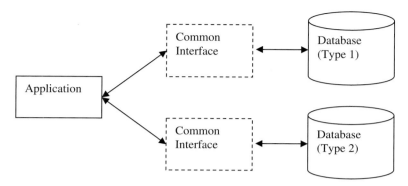

Fig. 7.2 Common interface for homogenising data sources

7.4.2.1 Purpose

Sometimes a particular application requires access to multiple data sources which have different formats. In this case it often makes sense to provide a common way of accessing the different types of data source. This can be achieved by adding a layer on top of the data source which hides the heterogeneity of the data sources from the application. Key benefits for business are reductions in development costs due to simpler applications and ease of integration of new data sources.

7.4.2.2 Implementation

Existing Solutions

In relational databases, SQL views are a common way of associating the results of an SQL query with a name. This name can be used as a short-hand to represent complex query results. This allows multiple tables within the same database to appear to a user as a single table. It also allows a table to be presented as having a different schema.

Query Translator Design Pattern

The Query Translator Design Pattern is a mechanism for accessing multiple heterogeneous databases in the same way.

Data is requested through a general query in a query language. The interface then gets the query translator to translate that query into the specific query which the data source will understand. It asks the data source for the data using the translated query, which the data source should provide. It then translates that data into the format that the interface between it and the application requires before returning the generic response to the application. In this way the same query can be sent to multiple data sources.

Component Development

OGSA-DAI contains a component which provides an implementation of SQL views. This SQL views component defines a mapping from view names to SQL queries representing these views, much as is done in relational databases.

SQL queries passed to OGSA-DAI are parsed and references to any view name replaced with the definition of that view. The parsed query is then passed to the database.

Although relational databases support this functionality, to define a view requires write access within the database of interest. However, in distributed or Grid environments, or in the use of publicly-accessible databases, users may only have read access to the databases of interest. OGSA-DAI's SQL views allow these clients to define views on top of these read-only databases.

The OGSA-DAI SQL views component was developed in conjunction with the BEinGRID project and version 1.0, compatible with OGSA-DAI 3.0 and 3.1, was released in December 2008.

Outstanding Issues

There are still open questions on how this capability can be realised for more diverse sources of data. If data exists in files and databases, producing a consistent query language becomes more difficult. This is also true if the data is in xml and SQL databases. Work has been done to provide XQuery interfaces to SQL databases [6] but a single consistent query language for all data is still not available.

7.4.2.3 Example of Use

BE12 Sales Management System [1] requires access to two different types of databases. This is a good example of a case where homogenisation of the presentation of the data makes a lot of sense. Without a level of abstraction, the rest of the system has to perform checks on the type of data source it is trying to access, making the final system more error-prone. In BE12 the system involved requires pizza shops to make information about their shop available to the pizza chain central office. In this case, it makes sense for the shops to keep a copy of the data, and provide the central office with a homogenised means of accessing it.

7.4.3 Synchronise Multiple Data Sources

Another important capability that is required by some of the BEs is to maintain the same information in multiple different locations. The goal is to have all of these sources containing the same data. Having this allows more users to access the data

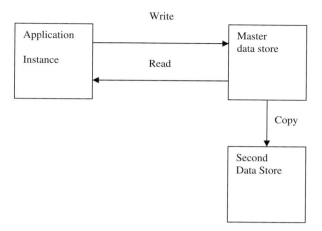

Fig. 7.3 Synchronising multiple data sources

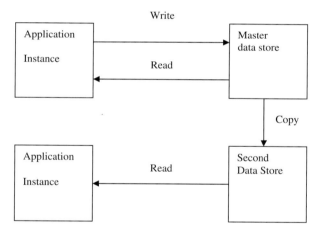

Fig. 7.4 Improving performance of read access using replication

while reducing the risk of a bottleneck occurring and it also provides redundancy if one of the systems goes down. It also allows multiple different companies to share information, reducing the amount of work required to enter information into both systems.

The simplest scenario is where one data location is set up to allow read/write access to the data. A second data store is then used as a backup. It performs all the operations of the first data store, but only takes over for read and write operations from the user if the primary fails.

To extend this solution to provide performance scaling, one can allow the second data store to act as a read only database. Even with this simple addition though, it is a challenge to keep information synchronised.

The problem becomes harder if the requirement is to replicate write access across multiple data sources, in which case all data sources in effect become masters (they communicate their updates to other data sources) and replicas (they reflect the changes made to other data sources). In that case, the access problems can be simplified by splitting up the data into distinct chunks that only one data source can alter. By doing this different parts of the data can be accessed and altered without having to copy information between the different data sources.

7.4.3.1 Purpose

There are a variety of reasons to want to synchronise the data held in multiple data sources. An important one is to improve the fault tolerance and thereby improve the quality of the service provided. This also relates to the business case for using this common capability, since having a robust system makes sense for any business. Another important use of synchronising data sources is to provide a reliable backup. This is very important for disaster recovery and those business cases where it is important to maintain the integrity of data. Another important reason is to reduce the amount of information that needs to be transferred manually between systems.

7.4.3.2 Implementation

Existing Solutions

DiGS [3] supports replication of data at the level of whole files, or groups of files, across two or more stores. The mechanism is intended to improve availability of data through redundancy. It also reduces volume of manual data transfers through data migration to particular stores based on prior user behaviour.

DiGS mitigates the synchronisation issue noted above by requiring applications to use a Replica Location Service in order to locate files. Specifically, an application sends a request for a file to a DiGS server, which then returns a list of available file locations, from which the application may choose the most suitable.

Moving our attention to database management systems, many already offer data replication mechanisms. These are typically available through configuration settings within the database engine. In addition, solutions like the High Volume Replicator [5] provide third party solutions which can also transfer information between databases in different locations and of different types. All of these solutions are focused on maintaining backups for disaster recovery.

If we consider a more general problem of responding to changes in a database there are a few other alternatives. SQL Triggers provide a mechanism to execute SQL statements whenever data in a database table changes.

Primary–Secondary Replicator Design Pattern

The underlying idea is identifying a change in one database and reacting to it. This very generic pattern describes an event-based reaction to a change in data. As with the data source publisher, this pattern also allows for an interaction with an existing system. Actions which affect the data source can be monitored and actions taken which communicate with other remote systems.

This pattern allows a backup of a data source to be prepared and made available if the primary data source fails. This allows a more robust system; if one machine goes down for some reason, other machines can continue to provide a service.

Component Development

The Primary–Secondary Replicator pattern was defined in the analysis of the BEs. Replication is already handled natively inside many relational databases. There are limitations however when trying to move information between databases developed by different vendors. The more general idea behind replication is reacting to a change in a database and performing an action that affects something else (another database for example). The OGSA-DAI Trigger component enhances OGSA-DAI by providing a mechanism for an OGSA-DAI workflow to be executed when a database is modified.

By providing a general mechanism for reacting to a change in a database, the OGSA-DAI Trigger component allows all the database access, transformation, and data delivery activities of OGSA-DAI to be used in response to a database change. An OGSA-DAI workflow can be executed automatically whenever a relational database changes.

This component is being used to interface with different B2B applications in BE24 (GRID2(B2B)). It provides access to a flexible Grid middleware for data management and a mechanism which allows the B2B extension to react to changes in the legacy system.

More information on the OGSA-DAI Trigger is available on Gridipedia [4].

Outstanding Issues

In terms of moving data between databases the most challenging scenario is one where multiple databases require both read and write access. In this case the problem of how to resolve conflicts when both databases change at the same time is the most interesting.

The first version the OGSA-DAI Trigger supports MySQL databases. The next release, due in quarter three of 2009, extends it to DB2, Informix and Microsoft SQL Server.

7.4.3.3 Example of Use

GRID2(B2B) allows B2B platforms to evolve significantly from the current state-of-the art. Currently data and process synchronisation (excluding expensive solutions deployed across B2B partners) between the participants of a B2B network require a human operator logging in to a portal or generating and processing files that represent supply-chain activities. What is missing is an affordable B2B platform extension to automate this synchronisation. While bigger companies can adopt new software, SMEs can only afford synchronisation if they can retain their original (legacy) infrastructure.

MaNeM [9] is one platform than can be used to manage the flow of information between partners in a B2B network. Different legacy software exists to perform supply-chain operations inside each company. The B2B platform provides work-flows which manage the interaction between the companies but data exists in parallel in the legacy systems and the same information has to be entered twice. This is achieved either through the use of a custom script which is run manually or by data entry by an employee at the company.

The goal of GRID2(B2B) was to produce a standalone extension to B2B platforms (not just MaNeM, but others as well). This extension allows information changes in one system to automatically update the B2B platform and other legacy systems. It achieves this through extensive use of the OGSA-DAI Trigger component.

7.5 Conclusion

This section describes some of the experiences which have been recorded through the interactions with the BEs. As the BEs progressed, their initial assumptions and intentions were clarified to better support their aims. This led to changes in the way they used data management ideas and components. The timescales of the experiments limited the complexity of tasks they could tackle, and this in turn limited the complexity of the scenarios that could be addressed by the BEs. Since our analysis was based on these experiments, this limited the scope of the problems we could address. Still, it is instructive to look at the experiments and try to draw some conclusion.

7.5.1 Recommendations

7.5.1.1 Do the Simple Thing if Possible

OGSA-DAI is intended for data access and integration scenarios. One of its strengths lies in the pre-written classes for accessing data from and delivering it to a variety of locations along with the ability to integrate custom and existing transformations into a workflow with ease. If an application only requires simple access to data, it may be easier to use a lower level library such as JDBC.

7.5.1.2 Use Native Database Functionality where Possible

There was initial interest from a number of experiments in scenarios where data was replicated between different databases. In one case, this requirement came from a desire to have a very robust system, which would switch to an up-to-date backup system, if the primary system failed. In this case there were no restrictions on the type of database to use. In this type of situation it is better to do as much as possible within the database if the database supports it. It is more efficient than deploying an additional layer of complexity in the form of middleware.

If there are many companies involved that employ different database management systems, using the native replication mechanisms might not be possible. In this case it makes sense to look for a solution such as OGSA-DAI, which provides a means of handling such situations.

7.5.1.3 For Simple Deployment of OGSA-DAI Data Resources Use the Data Source Publisher

To reduce the complexity for users new to using Grid middleware, use the Data Source Publisher. It reduces the amount of effort required to publish data via OGSA-DAI.

In particular, if you have data integration requirements coupled with developers who are not technical experts or are too busy, the combination of OGSA-DAI and the Data Source Publisher is a powerful tool for simplifying the process.

7.5.1.4 For Reflecting Data Changes Outside of a Database Use the OGSA-DAI Trigger

If the scenario is such that information in one database needs to be reflected outside that database, the OGSA-DAI Trigger provides a mechanism to perform actions in response to data changes within a database. It has the added benefit of making the OGSA-DAI workflows along with the data transformation processing and delivery mechanisms of OGSA-DAI available.

References

1. Business Experiment 12 (BE12) Sales Management System, http://www.beingrid.eu/be12.html
2. Business Experiment 24 (BE24) Grid technologies for affordable data synchronization and SME integration within B2B networks GRID2(B2B), http://www.beingrid.eu/be24.html
3. DiGS—Distributed Grid Storage (2009), Accessed 15 July 2009, http://www.gridipedia.eu/digs.html
4. Gridipedia—Gridipedia Technical Article—Data Source Publisher Pattern, http://www.gridipedia.eu/datasourcepublisherpattern.html

5. High Volume Replicator, http://www.highvolumereplicator.com/
6. Introduction to XQuery in SQL Server 2005 (2009), Accessed 15 April 2009, http://msdn. microsoft.com/en-us/library/ms345122.aspx
7. iRODS (2009), Accessed 15 April 2009, https://www.irods.org/
8. Java SE TEchnologies—Database (2009), Accessed 15 April 2009, http://java.sun.com/javase/ technologies/database/
9. MaNeM, Accessed 15 April 2009, http://www.joinet.eu/supply-chain/prodottosoluzioni.cfm? wid_cat=14&wid_pro=8
10. OGSA-DAI (2009), Accessed 14 April 2009, http://www.ogsadai.org.uk
11. OMII-UK (2009), Accessed 14 April 2009, http://www.omii.ac.uk/
12. SRB (2008), Accessed 15 April 2009, http://www.sdsc.edu/srb/index.php/
13. SwisSQL (2009), Accessed 15 April 2009, http://www.swissql.com/
14. Toyota Motor Manufacturing Kentucky, Toyota Production System Terms (2006), Toyota Motor Manufacturing Kentucky, Accessed 14 April 2009, http://toyotageorgetown.com/ terms.asp
15. Wikipedia (2009), Accessed 15 April 2009, http://en.wikipedia.org/wiki/Grid_computing

Chapter 8
Portals for Service Oriented Infrastructures

Efstathios Karanastasis, Theodora Varvarigou,
and Piotr Grabowski

Abstract Grid portals enable collaborative environments aiming to provide simple
and common Web interfaces to heterogeneous Grid resources and services. However, special factors must be taken into consideration when creating portal applications for business environments. This chapter discusses the approach taken by the
Portals technical area of the BEinGRID project, which resulted in the implementation of four software components that address security, user management, file management and management of computational jobs through Grid portals. The components, which were integrated in the Vine Toolkit framework—a collection of Java
libraries and User Interfaces for developing Grid applications, are characterised by
innovative features that aim to promote the overall business processes and comprise
an important improvement towards the business adoption of the Grid. The chapter
discusses in detail the technical and business aspects of the components and presents
examples of their usage in commercial environments.

8.1 Introduction

The Grid [4] is evolving from a tool of the research community to a powerful means
of improving business processes and increasing profitability. As Grid advances and
becomes more widely used at the commercial and industrial sectors, the need for
Grid environments supporting multi-user applications grows. However, the distributed nature of the Grid raises many concerns regarding ease of use, efficiency and
security. Addressing these issues is the key motivation behind the creation of Grid
portal applications targeting commercial sectors.

Grid portals enable collaborative environments aiming to provide simple and
common Web interfaces to heterogeneous Grid resources and services. Due to the
larger and wider user base they are intended for, commercial Grid portals should
simplify administration and problem solving by incorporating mechanisms for controlling access and performing administrative tasks in an efficient manner. Additionally, the portal should enable complex collaborations among systems with different
access control and security policies, and users with diverse authorisation levels or
different expertise. The latter is a very important factor, since business users are

E. Karanastasis (✉)
National Technical University of Athens, 9 Iroon Polytechniou, Zografou 15773, Greece
e-mail: karanastasis@telecom.ntua.gr

T. Dimitrakos et al. (eds.), *Service Oriented Infrastructures*
and Cloud Service Platforms for the Enterprise,
DOI 10.1007/978-3-642-04086-3_8, © Springer-Verlag Berlin Heidelberg 2010

typically not familiar with the Grid and its characteristics. The portal should aim at presenting a user interface which hides the Grid from the average end user.

This chapter presents the main outcome of the BEinGRID Portals technical area, which is based on the technical and business needs of eleven (11) Business Experiments (BEs) from the aerospace, architectural, financial, environmental engineering, automotive, pharmaceutical, textile, chemistry, IT and geological sectors.

The requirements of those BEs were elicited, examined and detailed. Their importance was evaluated according to different criteria, giving substantial weight to the business value, the technical innovation and the dependencies of each requirement. Thereafter, a number of common capabilities were identified. Common capabilities represent the refined common functionality needed by the Business Experiments in a number of different topics related to Grid portals, and they served as a basis for the provision of design patterns presenting an exemplar solution for each topic. Technical designs were provided for the following common capabilities identified in the area of Grid portals: Portals Security, User Management, Accounting, File Management, Database Access, Job Submission Monitoring and Control, Job Visualisation. Four of these common capabilities were implemented by following our design patterns. The outcome was a set of components that represent viable solutions to address the BE needs and their requested functionality and, at the same time, can be adapted to the various commercial Grid environments. At all times, our work has been driven by a main goal; Grid portals application in real-world business problems.

8.2 The Overall Challenge

The main challenge within the Portals technical area was to design and develop Grid portals components that enable user-friendly, low-complexity common Web interfaces enabling the transparent access and management of Grid services and resources aggregated from different distributed and heterogeneous sources. Hence, when developing the portal components it was utterly important to work on both levels—the user interface and the underlying business logic—with the same care. The user interface is the means for user interaction with the Grid and should be simple and user-friendly, but at the same time allow for the exploitation of the full potential of the underlying business logic. The business logic is responsible for the communication with the underlying heterogeneous services, and serves the user interface with the appropriate information to be displayed on the various portal pages.

The individual requirements and needs of the Business Experiments were taken into account during all phases of the portal components development. Nevertheless, the following special issues comprised the main challenges which led to important decisions:

• The business nature of the portal applications
• The variety of portal frameworks used
• The variety of Grid middleware used.

The business nature of the portal environment poses strict security requirements. Consequently, we had to cover the portal security needs by developing advanced mechanisms and tools suitable for use in business environments, and make sure that they could interoperate with all Portals components, as needed. This led to the idea of incorporating the various components in a common framework in which they would be able to interoperate seamlessly.

Additionally, the portal applications most of the BEs wished to implement intended to serve a rather large user base. Within this context, we should design tools for enabling the more efficient user management. Those users had in average less technical experience than traditional users of the Grid, i.e. scientific communities. Therefore, there was an increased need to design user interfaces that would be as user-friendly as possible and would utilise usage schemes the average commercial user is familiar with. Ultimately, the complexity of the Grid should be totally hidden from the end users.

Due to the variety of Grid middleware the BEs wished to use, it made sense to utilise an architecture with reusable common components and middleware-specific plug-ins that would provide the appropriate mappings and enable access to heterogeneous systems. This architecture would also allow for easy extendibility of the portal application with the usage of new plug-ins and modules for the connection to additional external systems and services. At the same time, the user interfaces had to be pluggable in different portal frameworks.

The development of the portals components was carried out at both the business logic and user interface levels. The components were implemented as inherent parts of the Vine Toolkit framework[1] [16]. In fact, the design of the Vine Toolkit was driven by the explicit needs of the BEinGRID BEs. The implemented solutions consist of a number of Java libraries complemented with a collection of Web 2.0 user interfaces, which represent the basic characteristics of the business-logic. Within the framework of the Vine Toolkit, the developed components can be used separately or interoperate in a seamless manner.

8.3 Common Capabilities and Technical Requirements

We present both the technical requirements and common capabilities of the Portals technical area in the same sections, grouped as per common capability. This approach prevents repetition of content and is easier for the reader to follow, since each common capability is closely related to specific technical requirements and

---

[1]The Vine Toolkit is a Java-based framework that offers developers an easy-to-use, high-level Application Programming Interface (API) for Grid-enabling applications. More importantly, Vine is a general library that can be deployed for use in desktop, Java Web Start, Java Servlet 2.3 and Java Portlet 1.0 environments with very little effort on behalf of the application programmer. Vine was designed to provide all the necessary entry points for "attaching" it to Web portals, Web Services and other "container" environments.

business needs. For each common capability we present an overview of the architecture and provide specific details of the implemented solution. Then, the business benefits are discussed and the usage is illustrated with a short scenario.

8.3.1 User Management

It is apparent that user requirements drive the development of portals and, at the same time, there is no need for developing a portal, if there is no user base to serve. In collaborative environments, the portal usually serves users of different roles within an organisation and consequently with different needs and rights within the portal. It is thus of fundamental importance that every portal application provides tools for the management of user identities, their access rights to content and resources and the environment presented to the users. Efficient user management also promotes the overall security levels of a portal application. Thus, User Management is closely related with Portals Security.

8.3.1.1 Requirements and Features

All eleven (11) BEs interested in implementing a Grid portal presented requirements related to user management. The needs identified mainly regarded viewing and editing personal information of users, and managing user accounts and user groups and their access to content or resources.

In the initial phase of requirements gathering we also identified that the two (2) most popular portal frameworks to be used by almost the all the BEs were, in order of popularity, GridSphere and Enginframe. Nevertheless, although Enginframe offered complete access rights management functionality, GridSphere was lacking these extended but important features at the phase of requirements gathering.

In summary, the main challenge for the Portals technical area was to identify or develop common tools that minimise the amount of effort for:

- Portal users to edit and store their personal information.
- Authorised portal users to view information of other users.
- Administrators to manage user accounts, their mappings to external accounts and their access rights.

The functionality provided by User Management is used by other portal components whenever there is the need to access or modify user information, or to check the permissions for access to a specific portal page or resource.

8.3.1.2 Architecture

Figure 8.1 presents the high-level logic of the User Management common capability in terms of user-management-related operations, which aims at fully covering the aforementioned functionality.

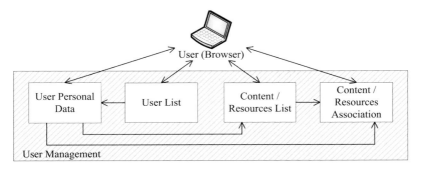

Fig. 8.1 User Management high-level architecture

The main operations visible on the figure are as follows:

- User Personal Data: Allows users to view and change their personal information, like first and last name or email address. The information is retrieved from the user database and when modified is stored back to the database.
- User List: Allows users or administrators to view or manage (e.g. add, delete) accounts of portal users. This also interacts with the user database to update its records.
- Content/Resources List: Presents all the available portal resources, like portal pages, users could possibly be assigned access to.
- Content/Resources Association: Covers the administrative need to change the association between user accounts and content or resources. For example, the administrator can choose which portal pages are available to users and their access rights, according to their user group.

This design also addresses operations related to controlling access rights of users and groups. For the shake of completeness of the design, we decided to present this initial version, which was created right after the identification of the User Management common capability.

8.3.1.3 Implementation

The User Management design presented in Sect. 8.3.1.2 covers the full set of features, as discussed in Sect. 8.3.1.1. As also mentioned in the latter section, the most popular (due to a number of reasons) portal framework the BEs intended to use, GridSphere, initially lacked basic user management features which are of significant importance in a business environment. However, version 3.0 of GridSphere was out before the development of the User Management component started. The new version of GridSphere included a native Role Based Access Control (RBAC) system, which rendered it obsolete to implement our initial design in its full extent.

Due to this fact, in the development phase of BEinGRID we made the decision to:

- Implement a generic tool for managing user data with the capability to use external systems for assigning user roles and managing their access rights, and additionally.
- Implement a module that would allow our component to connect to the Role Based Access Control (RBAC) system of GridSphere and take advantage of the offered functionality.

Thus the "User Personal Data" and "User List" operations of the design where implemented inside the Vine Toolkit, while the "Content/Resources List" and "Content/Resources Association" where not implemented, due to the fact that the GridSphere team had already addressed the issue of GridSphere lacking this functionality. In the following paragraphs we elaborate on this specific configuration (i.e. User Management component + GridSphere RBAC), but the reader should keep in mind that it is possible to use another external role management system with the User Management component. This architecture improves flexibility by allowing usage of the most appropriate role management system/portal container for each case.

In typical portal applications, user data used to be kept in the database of the portal container. User Management is intended to replace that database by the Vine Toolkit user database, which is now used to maintain the basic user information. Since the RBAC system of the portal container is used, the container's database is still used for maintaining information related to the group(s) each user belongs to and information regarding the access rights of each group. One Vine user account is always associated (mapped) to one and only one user account at the portal container. Both accounts are created and maintained automatically by User Management with the aid of Portals Security (see Sect. 8.3.2), which is used to perform all registration, authentication and authorisation actions requested by the User Management component. Apart from the container account, more external accounts can be associated with a Vine user account (see Sect. 8.3.2.3).

The User Management component also provides new Web interfaces for interaction with the Vine user database. The interfaces allow new users to sign up, existing users to edit their personal data or view detailed information of other users, if authorised, and administrators to oversee and maintain user information and account mappings.

8.3.1.4 Innovation and Business Value

User Management keeps account of the portal user details and the mappings of their identities with external ones in an innovative manner (also see Sect. 8.3.2.4 about Portals Security), allowing the utilisation of a highly flexible system, where users can be granted access to new Grid services immediately. In addition, it allows using an external system to control user access to specific services and data according to assigned access rights. Thus, it helps improve the overall levels of system security in an effective and robust manner and is of fundamental business importance in every portal application.

The provided user interfaces make it and efficient (fast and simple) for businesses to manage their portal user accounts, where portal users may be employees of the companies involved in a particular business, customers or the general public.

Another benefit lies in the fact that these important features offered by the User Management component can be exploited in various configurations, in conjunction with a number of different access control systems and portal containers, for successfully covering different business needs.

As already mentioned, User Management uses the capabilities of Portals Security to carry out important operations. Thus, the reader is advised to also refer to Sect. 8.3.2.4 for an overview of the benefits that the two components can offer.

8.3.1.5 Example Usage

The reader should refer to Sect. 8.3.2.5 for a joint usage scenario of User Management and Portals Security.

For additional detailed scenarios, please refer to [11] and [13].

8.3.2 Portals Security

Offering Web-based access and management of resources and service capabilities poses strong security requirements. The security threats in business environments are numerous and their consequences range from plain user frustration to serious security attacks with apparent social, economic, legal or organisational impacts.

Most security-related issues are covered by the "Security" technical area of BE-inGRID (see Chap. 4). This section only aims at covering the portal-specific needs analysed below.

8.3.2.1 Requirements and Features

As one would expect given the importance of the security topic, the total of the BEs that aimed at including a portal in their business solution had security requirements. Depending on the use case to be considered, the requirements of these eleven (11) BEs were mainly related to login/logout procedures, access to content hosted within a portal, access to external content or resources made accessible from a portal and integration with third-party security services. It is obvious that some of these requirements are also related to User Management, discussed in Sect. 8.3.1.

While working to tackle these requirements, we identified another functionality that would significantly ease the management of processes related to authentication, authorisation and integration of third-party security systems, as well as improve the overall levels of security in a portal application. This functionality is related to user registration in the portal and in underlying services and Grid middleware accessible through the portal. To address these issues we created an innovative mechanism that we call Single Sign-Up, which is described in more detail below.

Based on these needs and given the variety of Grid middleware and portal frameworks used in the various BE environments, the main challenge for the Portals technical area was to identify and develop common tools that minimise the amount of effort for portal developers and administrators to:

- Register users in the portal and provide Single-Sign-Up registration to heterogeneous external resources made accessible by the portal for its users.
- Authenticate users in the portal and provide Single Sign-On authentication to heterogeneous external resources.
- Integrate third-party security services at the portal user interface level and configure their usage. The reader should note that Portals Security is not concerned with how actual third-party services are implemented but rather how they can be incorporated and utilised in portals.
- Aid authorisation of user access in content and resources exposed from within the portal, by using third-party systems.

8.3.2.2 Architecture

Figure 8.2 presents the high-level logic of the Portals Security common capability in terms of security-related operations, as designed to address the aforementioned functionality.

The proposed architecture covers the following main operations:

- Account Creation (Single Sign-Up): Allows automated, user-friendly user registration at the portal and to underlying Grid middleware and third-party services used through the portal.

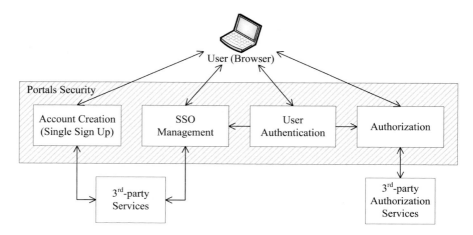

Fig. 8.2 Portals Security high-level architecture

- User Authentication: Allows the user to login into the portal and be authenticated.
- SSO Management: Allows automated Single-Sign-On user authentication in various Grid middleware and services used through the portal. The user must have been previously registered in these external services. Different authentication schemes are supported.
- Authorisation: Allows the administrator to access third-party authorisation systems to change user permissions. Different authorisation schemes are supported.

8.3.2.3 Implementation

As mentioned above, the Portals Security and User Management components complement each other in the implementation of the Vine Toolkit. In addition, Portals Security offers its services to other components of the Grid portal, if they require access to remote third-party services or middleware in a secure way.

While User Management is mainly concerned with maintaining the user database and providing the appropriate user interfaces for editing its contents, Portals Security deals with the underlying mechanisms for registration and authorisation in third-party services, and management of the associated credentials. It offers a number of interfaces covering a wide range of the security needs of a typical Grid portal application, enables Single Sign-On and promotes the innovative concept of Single Sign-Up. Single Sign-Up allows, depending on the configuration, automatic creation of user accounts and registration in a number of chosen third-party services (such as Grid middleware platforms) during sign-up or after account approval, simplifying the process of generating credentials and registering them with Grid middleware, or creating accounts on remote systems. In specific, the following Grid platforms are currently supported: Globus Toolkit 4 [5], UNICORE 6 [3] and gLite 3 [2].

Specifically, Portals Security is used to register Vine Toolkit users as required by the User Management component (for more information, please refer to Sect. 8.3.1.3). Users are typically registered in the portal container and in every Grid middleware they are supposed to have access to. This is defined for the users of a specific domain in the configuration of the Vine Toolkit. The Portals Security component makes it possible to organise resources into a hierarchy of domains to represent one or more Virtual Organisations (VOs) [6]. Portals Security creates the required registrations and the corresponding credentials, using its internal or an external Certification Authority (CA) to sign them. It then stores them in the component's internal credential repository. These external registrations are mapped to a Vine user by the User Management component. This innovative registration mechanism is called Single Sign-Up.

Each time a registered user logs in to the portal, related security information for that user (username, password, etc.) is retrieved and stored as session parameters or proxy certificates in the internal repository and used to also authenticate that user at the portal container. Then, e.g. during File Management (see Sect. 8.3.3) and Job Submission (see Sect. 8.3.4) operations, this information is passed to the external services, as appropriate according to the authentication method they utilise. This provides an advanced Single Sign-On solution.

Portals Security also provides complete credential management functionality and the associated user interfaces, and the ability to use an external credential repository, like MyProxy, for optionally retrieving existing credentials, if so wished.

For user authorisation, Portals Security allows connection to third-party authorisation systems. A combination of more than one authorisation systems can be used as required, e.g. an Access Control List (ACL-based) system and a Role Based Access Control (RBAC) system, which could be integrated in the portal container.

8.3.2.4 Innovation and Business Value

Portals Security and User Management are able to support and orchestrate Web based registration and authentication to the Grid. By using the innovative Single Sign-Up concept, administrators no longer have to undertake new user registrations at the portal manually. Accounts on underlying remote systems and services, usually deployed on different machines possibly utilising different middleware technologies, can also be created and maintained automatically. At the same time, the associated credentials can be created, signed, stored and delegated as required with no human intervention. This has a positive impact on the efficiency of the processes and the responsiveness of the system. Not only administrative effort is reduced, but new users can be granted access to the system rapidly, which results to increased productivity and reduced cycle time, but also denotes end user confidence to the system.

The flexibility in creating and managing user accounts can also be exploited in favour of the system's security. Instead of adopting a many-to-one portal-to-Grid identity mapping, each user account at the portal can be easily mapped to a unique user account at the remote service's side, which satisfies strict security requirements and realises a highly flexible system. Additionally, the abstraction of administrators from certain procedures decreases the human error factor. At the same time, administrator expertise requirements can be lowered, which leads to reduction of administrative expenses.

With regard to end users, the details of complex security related operations, as well as the complicated architecture of the Grid, are hidden away from them. Moreover, by making use of User Management and Portals Security, different external security systems can be integrated easily and uniformly in the portal, as required for accessing heterogeneous Grid resources or legacy systems, thus allowing preservation of existing investments in technologies and knowledge.

8.3.2.5 Example Usage

This section presents the overview of a generic scenario utilising User Management and Portals Security for enabling a number of administrative capabilities applicable in various portal environments.

In the scenario, the portal administrator wants to set up the portal in a manner that it can accommodate a number of users from various roles. According to their role, it must be possible to organise users in different groups and specify each group's permissions and access rights to content on the portal. Additionally, each portal user must be able to interact only with specific Grid middleware, third party services or resources accessible through the portal. The administrator needs tools to easily register users to the Grid middleware and third party services, and to manage these registrations and the associated credentials.

For the full scenario, please refer to [11]. For additional detailed scenarios, please refer to [14].

8.3.3 File Management

Resources are an important part of the computational-Grid, but so is data. Humans or machines feed data into computational resources for the production of results. Then, the resulting data is retrieved and evaluated or inserted to another machine for further processing. In some cases, distributed data is the main reason for the existence of a so-called data-Grid. In order to process or maintain data, humans and/or machines require tools for moving it across platforms, handling and sharing it.

8.3.3.1 Requirements and Features

Seven (7) BEs presented direct file management requirements. Nevertheless all of the BEs indirectly expressed the need for file-management-related functionality, required for achieving the scopes of other portal capabilities, such as job submission (see Sect. 8.3.4). Depending on the specific business case, there were requirements to access the file system at the Grid middleware in order to upload files to be used as input for job submission and to download resulting output files after successful completion of job execution, to enable a personal storage space for each user to temporarily store files used while working at the portal, or to allow access to a collaborative repository, where users would be able to copy job results to share them with their colleagues.

The various BEs intended to make use of different Grid middleware and required to access file management services of different types via the portal, while in some cases the same BE required its users to access heterogeneous repositories through the same portal application.

Taking into account these facts and requirements, the main challenge for the Portals technical area was to identify and develop common tools that would allow easy and efficient execution of the following operations through a common user interface:

- Connect to a number of heterogeneous file repositories and file systems in a unified manner
- Create/delete folders

- Upload/download files
- View/access/delete files
- Change properties and access rights of files and folders
- Perform file transfers between Grid repositories of different types
- Enable a personal storage space for each user at the portal.

8.3.3.2 Architecture

Figure 8.3 presents the high-level logic of the File Management common capability in terms of file-management-related operations, as designed to address the afore-mentioned functionality.

File Management covers the following operations:

- Repository Selection: Allows the user to choose the Grid repository location to work with. This can be done by means of a direct URL, or by a list of available locations displayed at the portal. The history of previous user locations can be used, or a call-out to an external resource list or resource discovery service can be made, in order for the list of available locations to be prepared.
- Repository Management: Allows the user to perform a number of operations on files and folders residing in the chosen repository:
  - Folder Creation/Deletion: Create a new folder and name it, or delete an existing folder in the repository.
  - File/Folder Properties: Display the properties of the selected file(s) or folder(s) and change some values, if allowed by the remote file system.
  - File/Folder View: Allows users to view the contents of a folder or display the content of specific file types, such as text files. It also provides the Web browser

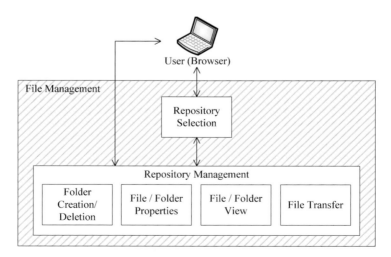

Fig. 8.3 File Management high-level architecture

application with the information needed to properly display or save a file in the filesystem of the user's local machine, according to its type.

- File Transfer: Allows the user to upload or download files to or from a specific location.

8.3.3.3 Implementation

File Management was implemented in the Vine Toolkit in accordance with the presented architecture. The underlying business logic allows to connect to file services of different Grid platforms or to external file repositories and to perform file transfers when specifying stage-in and -out operations at the Job Submission Monitoring and Control component (see Sect. 8.3.4.3). Currently, support for the following services and protocols is offered: GridFTP (e.g. for gLite and Globus Toolkit), Unicore 6 Storage Management Service (SMS), GRIA Data Service. The support of GRIA [15] was added within the context of the WOW2GREEN [17] Business Experiment of BEinGRID.

The user interface of the File Management component was designed in the well known form of existing file managers, users are familiar with. The main intention was to make it as easy and time saving as possible for new users to adapt to the environment of the Grid portal. Also, these designs have been used and evaluated for years, and have proven to be highly usable and widely accepted by end users. The user interface features two modes.

The "Browse" mode allows a single repository to be presented on the portal page. The user can choose between two different standard views, which can be switched by pressing the appropriate button. The "Icons" view displays large icons for each file or folder, while the "Detail" view presents a list of files or folders along with their detailed info, such as size and date of last modification.

The "Transfer" mode offers a two-panel view, which allows displaying two different file repositories side-by-side on the same page, making it very useful for file transfers. On the one side, the user can select a remote or local repository, and on the other side another repository. Then, the user can copy files from one repository to the other simply by dragging and dropping them using the mouse, or by using the appropriate buttons.

8.3.3.4 Innovation and Business Value

The innovation, and at the same time business value, of File Management mainly lies in both its business logic and user interface. Its ability to access repositories of different types through the portal enables improved management of data assets and improved reallocation of data resources in a simplified and uniform way. File Management makes it possible to use existing file repositories without the need for modifications, thus saving costs and speeding up the processes of setting up the Grid portal.

Web 2.0 technologies were utilised in the File Management user interface to provide a user-friendly portal environment. That user interface enables easy execution of advanced file management or file transfer operations by utilising graphical interaction methods average commercial users are already familiar with. Because of this fact, the users do not need to be excessively trained before being able to use the full capabilities of the Grid portal. Additionally, the familiar environment makes users better trust the system, and increases productivity and profitability by reducing the chances of user frustration and minimising the time needed to perform operations.

8.3.3.5 Example Usage

Various examples for the usage of the File Management component can be deduced from the above sections. However, here we present the overview of a usage scenario taken from the supply chain sector, where a portal is set up for enabling collaboration between users of different roles. The roles supported are Suppliers (companies producing the products), Distributors (warehouses) and Retailers (stores selling the products to the end users).

Among the other operations performed through the portal, the Suppliers are responsible for creating reports for the Distributors, which are uploaded and stored in a collaborative file repository. The Distributors download and view the Supplier reports, then create and upload reports for all their Retailer clients. Finally, the Retailers can access reports from the Distributors supplying them. Connection to the collaborative repository is achieved by means of the File Management component.

For the full scenario, please refer to [11]. For additional detailed scenarios, please refer to [10].

8.3.4 Job Submission Monitoring and Control

One of the fundamental uses of the Grid, as initiated by the research community several years ago, is the execution of computationally intensive jobs, such as simulations of different kinds. As Grid advances and becomes a powerful means of improving business processes and increasing profitability, the need for execution of computationally intensive jobs remains, but the necessity for simple environments grows. One of the roles of Grid portals is to enable commercial and industrial users to access the computational power of the Grid without the need to understand its architecture and complexity.

8.3.4.1 Requirements and Features

As expected, a large number of BEinGRID Business Experiments (BEs)—ten (10) in specific—presented requirements for managing computational jobs. These requirements were initially categorised into two different topics, namely "job submission" and "job monitoring and control", which were then merged into one due to their inseparable nature. The main business need in this area was to submit jobs for execution at the Grid middleware. Some BEs required that their users would be able to choose different computing resources or Service Providers depending on the applications to be run. Furthermore, there was the requirement to monitor jobs as they execute, i.e. view job status, and control jobs that have not finished, i.e. suspend and resume or cancel them.

At the phase of gathering requirements, we identified an additional feature that would considerably ease portal users and help organising the rest of the functionality. That feature was the ability to display the history of previously submitted jobs for each portal user.

Based on these requirements and given the variety of Grid middleware used by the BEs, the main challenge for the Portals technical area was to identify and develop common tools that allow portal users to:

- Select a Grid computing resource
- Submit computational jobs to heterogeneous Grid resources
- Monitor the status of executed jobs
- Control the execution of jobs
- Display job history of previously submitted jobs.

8.3.4.2 Architecture

Figure 8.4 presents the high-level logic of JSMC in terms of job-management-related operations, which aims to cover the aforementioned needs and requirements.

The JSMC component implements the following operations:

- Job History: Presents the history of submitted jobs, along with some submission details.
- Job Submission: Allows the user to define and submit a new job to the middleware. It consists of a number of internal operations, which can be performed in any order. After the submission of a job, the component automatically proceeds to the Job Monitoring operation.
  - Resource Specification: Allows the user to select the computing resource, where the selected job will be executed.
  - Application Specification: Allows the user to specify the application to be executed on the Grid and the various application parameters and arguments, including input and output files. The application must be already installed in the resource selected.
  - Requirements Specification: Allows the user to specify different requirements needed for the execution of the job, such as amount of memory, number of CPUs, etc.
  - Data Specification: Allows the specification of input and output staging, in order for the user to optionally define which files should be copied to and/or from the working directory on the host where the job will be executed before job submission and/or after job completion respectively. The business logic of File Management (see Sect. 8.3.3) is used to perform this operation.

Fig. 8.4 JSMC high-level
architecture

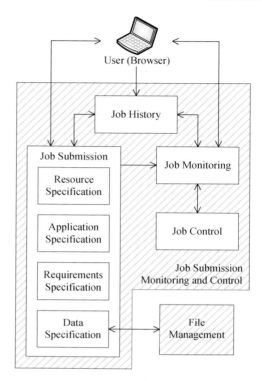

- Job Monitoring: Presents information about the submission and the state of a specific computational job.
- Job Control: Allows the control of a specific job, if it is still active. A running job can be paused or cancelled, while a paused job can be resumed or cancelled. The support of these operations depends on the Grid middleware in use.

8.3.4.3 Implementation

The business logic of Job Submission Monitoring and Control in the Vine Toolkit allows taking advantage of the resources offered by heterogeneous Grid platforms. The Grid middleware currently supported are: Globus Toolkit 4, UNICORE 6, gLite 3 and GRIA 5.3.

The Job Submission Description Language (JSDL) specification [7] is used to describe generic jobs. Before the job is actually submitted to the middleware plug-ins, the JSDL job description is translated into the appropriate formats, if needed. For the UNICORE 6 plugin, no translation is needed, since UNICORE 6 fully supports the JSDL specification. For Globus Toolkit, a translation to the GT 4.0 WS GRAM Job Description Schema is taking place. For gLite 3, the JSDL compliant job description is transformed into a Job Description Language (JDL) Attributes Specification compliant job description. In the case of GRIA 5.3, the translation process could be lossy, since GRIA only supports a sub-set of the JSDL tags. The non-supported tags are simply omitted. All translations of job descriptions from generic to native formats are made with the use of Xalan [1], which implements the XSL Transformations (XSLT) Version 1.0 [1] and XML Path Language (XPath) Version 1.0 [1] specifications.

Due to its generic nature, the user interface of the JSMC component is designed in the form of a "standard" and an "expert" view. The "standard" view consists of detailed pages (tabs), each focusing on a specific group of parameter types. The "expert" view allows expert users to write or change their own JSDL descriptions by editing the corresponding XML file. All changes made in the "standard" view are reflected immediately in the "expert" view and vice versa.

8.3.4.4 Innovation and Business Value

Job Submission Monitoring and Control is the basic functional element of every portal application designed for enabling user access to the computational Grid. JSMC allows businesses to expose a number of heterogeneous services through a common user interface. In case the infrastructure used is not in-house, JSMC makes it easy for businesses to choose between several Service Providers, even if they utilise different Grid platforms, and use their offered services according to the business's own interests. At the same time, JSMC allows easy incorporation of existing systems and Grid middleware. Compatibility with existing systems denotes reduced integration costs and preservation of investments in technologies and knowledge.

The ability of JSMC to store and display history and important details of previously submitted jobs could prove very useful in terms of financial management and control of all job transactions held through the portal. Job control helps minimising the usage of computational resources, e.g. when users cancel a recently submitted long-lasting job because they found out that they had input some incorrect parameters, which is directly associated with minimising the operational costs of a business.

Because of the common user interface, new user training is limited to just one user interface, regardless of the application or Service Provider/Grid platform they need to use, thus further minimising the costs related with the transition to a new portal environment for existing businesses. In addition, while making it considerably easier for novice users to justify their computational needs, the user interface of JSMC also offers advanced capabilities for experienced users, enabling exploitation of the full Grid potential. The improved usability of the system and the simplification of all procedures increase user satisfaction, reduce the chance of user frustration and increase productivity, while the job control functionalities give the portal users immediate control over the executed tasks, thus increasing user confidence in the system.

8.3.4.5 Example Usage

This section presents the overview of a simplified scenario based on the needs of a small enterprise from the engineering sector to execute computationally intensive jobs (e.g. simulations). The portal serves the few users of the corporation. The users prepare and submit computationally intensive simulations through the portal to the infrastructure provided by a number of Service Providers the corporation has an agreement with. Each Service Provider only supports specific simulation applications and utilises a specific Grid middleware platform. The portal provides a uniform interface to access these heterogeneous computational resources. In addition to submitting new simulations to the computational Grid, a portal user is able to view a list of previously submitted simulations and their status. The user can further pause and resume, as required, running simulations or cancel a simulation, if incorrect parameters were input at the job specification phase, for instance.

For the full scenario, please refer to [11]. For additional detailed scenarios, please refer to [12].

8.4 Conclusion

In this chapter we presented the main outcomes of the Portals technical area of the BEinGRID project. The User Management, Portals Security, File Management and Job Submission Monitoring and Control components were implemented as parts of the Vine Toolkit framework and were further refined in a constant interaction with the BEs. The Vine Toolkit is now in a mature state and available for download through the Gridipedia repository [9]. As a conclusion, we would like to present a lesson learnt while working on the components and some recommendations for portal developers.

As mentioned above, the Portals components are all part of the Vine Toolkit. Vine, as developed, is a very useful tool providing a wide palette of Grid functionalities for a wide spectrum of different middleware, making this both its advantage and drawback, since due to its size its maintenance comprises a rather hard task—it consists of about 200,000 lines of code in 2,000 Java files, not including Graphical User Interface and configuration files.

In addition, we would like to make a number of recommendations for portal designers and developers. First of all, prepare carefully your work-plan and allow enough time for testing and resolving problems. This phase may take longer than expected; even more than the initial implementation phase took. Some Business Experiments expressed too many initial requirements and expectations, but we found out that they were only able to implement a few of them in the given timeframe.

When considering a portal solution, do not reinvent the wheel. Look at well know existing portal frameworks and evaluate them according to your specific needs. Check:

- Which of them best suits the usage scenario of your business case.
- If it is feasible to build a portal implementing the needed functionality and the desired user interface with each of the available portal frameworks.
- How easy and time consuming it is to implement the required changes (if any).

After choosing the portal framework your main scope should be to make your solution as easy as possible—Grid and Grid middleware are complicated enough.

If you have the need for a Portal functionality identified by the portals technical area (for a list of identified functionality topics please refer to Sect. 8.1) but not developed within BEinGRID we recommend following the architecture described in our early design patterns [8]. We have provided designs for all the identified topics and believe that our designs illustrate the best way of addressing the specific Grid portal issues and requirements.

Finally, we suggest that you include in your Gird portal solution features such as Portals Security and User Management, which are of grave importance in any business environment.

References

1. Apache Xalan, XSLT and XPath, http://xml.apache.org/xalan-j/
2. A. Edlund, Programming the grid with gLite. Computational Methods in Science and Technology **12**(1), 33–45 (2006)
3. D.W. Erwin, D.F. Snelling, UNICORE: a grid computing environment, in Proceedings of the 7th International Euro-Par Conference on Parallel Processing, Manchester (Springer, Berlin, 2001)
4. I. Foster, What is the Grid? A Three Point Checklist, GRIDToday, Jul 2002
5. I. Foster, Globus Toolkit version 4: software for service-oriented systems, in Proceedings of the IFIP International Conference on Network and Parallel Computing, Beijing, China (Springer, Berlin, 2006)
6. I. Foster et al., The anatomy of the grid: enabling scalable virtual organisations. International Journal of Supercomputer Applications **5**(3), 200–222 (2001)
7. Global Grid Forum, Job Submission Description Language (JSDL) Specification, Version 1.0, Nov. 2005, http://www.ogf.org/documents/GFD.56.pdf
8. Gridipedia Technical Article—Portals, http://www.gridipedia.eu/portals-article.html
9. Gridipedia—The European Grid Marketplace, http://www.gridipedia.eu/
10. E. Karanastasis, P. Grabowski, Grid Portal solutions—File Management, http://www.gridipedia.eu/technicalwhitepapers.html
11. E. Karanastasis, P. Grabowski, Grid Portal solutions for Business Environments, http://www.gridipedia.eu/fileadmin/gridipedia/po_cluster/Whitepaper_Grid_Portal_solutions_for_Business_Environments.pdf
12. E. Karanastasis, P. Grabowski, Grid Portal solutions—Job Submission Monitoring and Control, http://www.gridipedia.eu/technicalwhitepapers.html
13. E. Karanastasis, P. Grabowski, Grid Portal solutions—User Management, http://www.gridipedia.eu/technicalwhitepapers.html
14. E. Karanastasis, P. Grabowski, Grid Portal solutions—Security, http://www.gridipedia.eu/technicalwhitepapers.html
15. M. Surridge, S. Taylor, D.D. Roure, E. Zaluska, Experiences with GRIA Industrial Applications on a Web services grid, in Proceedings of the First International Conference on e-Science and Grid Computing (IEEE Computer Society, Los Alamitos, 2005)
16. The Vine Toolkit framework, http://vinetoolkit.org
17. WOW2GREEN Business Experiment, http://www.beingrid.eu/wow2green.html

Chapter 9
Bringing it all Together

**Angelo Gaeta, Theo Dimitrakos, David Brossard, Robert Piotter,
Horst Schwichtenberg, André Gemünd, Efstathios Karanastasis,
Igor Rosenberg, Ana Maria Juan Ferrer, and Craig Thomson**

Abstract In this chapter we first summarise the business challenges the innovation opportunities in each thematic area (Sect. 9.1). Then we explain the dependences between the common technical requirements in each area (Sect. 9.2). Afterwards we summarise the common capabilities developed by the BEinGRID programme in order to address these opportunities (Sect. 9.3). Section 9.4 presents examples of scenarios where a large number of these innovations are brought together in order to solve a complex problem. The focus of this section is to stress the "plug-n-play" approach allowed by the BEinGRID Common Capabilities, validating it through integration scenarios, that demonstrates how identified capabilities are combined.

9.1 Business Benefits and Innovations per Thematic Area

This section summarise the business challenges the innovation opportunities in each thematic area.

9.1.1 VO Management

The activities of the VO Management area have led to the identification of Technical Requirements, Common Capabilities, Design Patterns and Software components to address the issues of governance and lifecycle management of a VO, including aspects of security and semantics in a VO.

Details on the activities and results of the VO Management area can be found in Chap. 3. The purpose of this section is to give an overview of the business benefits and innovation of the VO Management results.

The main challenges addressed by this area are the creation and management of a secure federated business environment among autonomous administrative domains, the separation of concerns between provision and management of application services and operational management of the VO infrastructure (e.g. separating the coordination of application execution from Resource monitoring), and the automatic

A. Gaeta (✉)
Centro di Ricerca in Matematica Pura ed Applicata (CRMPA) c/o DIIMA, via Ponte don Melillo,
84084, Fisciano (SA), Italy
e-mail: agaeta@crmpa.unisa.it

T. Dimitrakos et al. (eds.), *Service Oriented Infrastructures
and Cloud Service Platforms for the Enterprise*,
DOI 10.1007/978-3-642-04086-3_9, © Springer-Verlag Berlin Heidelberg 2010

discovery of available resources or services which meet a given set of functional requirements inside a VO or among different VOs.

The results produced by the VO thematic area support and address the key problems reported above.

For instance, the VO Set-up component is a web service providing capabilities to support the VO Identification and Formation phases where members of the VO have to be identified and a circle of trust among them has to be created. The component allows for the management of VO related registries and a secure federation lifecycle.

In terms of business benefits, the VO Set-up component and the capabilities implemented allows for agility in responding to new needs/requirements and improved time-to-market (by set-up of a VO when a new opportunities arises); improved trust in Business to Business interactions, and dealing with the geographical and organizational distribution of teams and computational resources.

In terms of innovation, with respect to other solutions for VO management, the model of the VO Set-up is better suited to the way enterprises thrive nowadays where new opportunities rise and fall quickly and where the environment is very prone to change. The VO Set-up allows for more flexible, business-driven interactions. Trust is established from the VO Set-up through to the security components in particular the Security Token Service.

Our solution implements the model defined in the TrustCoM project [8] that is a distributed credential and policy based model allowing the establishment of asymmetric and binary trust relationships.

The second challenge of this area, namely the separation of concerns between application provision and VO Infrastructure operational management, is addressed by the Application Virtualization component. It is a web service providing functionalities to create business capabilities required for the Operational phase of the VO and configure infrastructural services for secure message exchange in VO and monitoring & evaluation of the Service Level Agreements.

The virtualized application is exposed via a Gateway and the configuration of infrastructure services (potentially provided by third parties) for managing non-functional aspects of the application is done in a transparent way for the application consumer. So, the added value is mainly in the automatic configuration of third party management services such as SLA and security. The adoption of the Gateway avoids direct access to the resources of a SP and access is controlled by the security services.

In terms of business impact, the Application Virtualization allows ASPs to expose their applications in a simple and manageable way without being involved in the management of the enabling infrastructure. This increases flexibility and allows a separation of concerns between application provision and management, and enables the transition towards a SaaS model.

Lastly, the Automatic Resource Discovery (see Chap. 3 for details) integrates a semantic layer on top of the GT4 Monitoring and Discovery System (MDS) [3] to improve the process of resource discovery in a VO. It allows reasoning over ontologies. Resource Discovery, obviously, can be done in several ways. The most popular one in Grid environments is the adoption of the GT4 MDS. The Automatic

Resource Discovery component, indeed, is built on GT4 MDS and it augments the MDS index service by providing a Query Service capable of executing SPARQL [7] queries.

In terms of business impact, this component allows for use of non dedicated resource on demand (so, reducing the TCO), and agility in responding to new needs and requirements (by discovering resources on demand).

The main advantages of using the component compared with services such as GT4 MDS is a simpler and less ad-hoc user interface, an extended information set, a common repository and interface for application-specific information and reasoning over an ontology rather than simple string matching of requirements against stored values.

Finally, it is worth mentioning that the results of the VO area can be enhanced and combined with other results of the BEinGRID technical activities. This aspect will be better described in the other sections of this document. We only anticipate that the combination of results coming from the VO, SLA and Security areas can help in addressing most of the challenges currently associated with ad-hoc dynamic collaborations.

With ad-hoc dynamic collaboration, we refer to the case in which the VO members have to be dynamically identified on the basis of the business goal of the VO. Of course, after the identification, policies and agreements have to be negotiated and, generally, there is no trust a priori among the partners, so trust & identity management is a key factor for the success of this kind of collaboration.

The ad-hoc dynamic collaboration presents several challenges such as partner identification and selection, agreement negotiation, establishment of secure federation among different administrative domains, set-up of the distributed infrastructure underpinning a VO, service creation, exposure and management, partner and services monitoring and replacement, and update of the infrastructure when context changes. Those challenges, as it will be better described in Chap. 3, can be addressed by adoption of complementary results form VO, Security and SLA technical areas.

9.1.2 SLA Management

This section has the purpose of giving an overview of the business benefits and innovation of the SLA Management results. Details on SLA Management in the BEinGRID project can be found in Chap. 5.

SLA Management builds on the definition of a machine readable document, the Service Level Agreement (SLA), that specifies the Quality of Service that a certain service offered by a provider is able to deliver to its consumers. This SLA not only describes the desired quality attributes, but also can define conditions that apply in case of a service failure, e.g. a financial compensation that would be given. The service delivery is described through the {service, SLA} pair, defining exactly what the client is expecting from the provider. The complete lifecycle of the service is mirrored by the life-cycle of the SLA. As such, the SLA is expected to be dynamic, as its lifespan is equal to the service usage by the client.

After being discovered, the service SLA is negotiated through the specification of individual service level objectives. Then, during the lifetime of the service the service quality, e.g. the response time of the service, is monitored and violations are automatically identified. This enables the service provider and also the customer to automatically act upon a service failure. To be effective, the whole SLA lifecycle must be seen as a management capability of the provider, and its minimal steps must be implemented. The benefits of the minimal steps are described in the paragraphs below.

The main challenge of SLA Negotiation resides in providing a comprehensive environment for offer and demand bargaining, in the legal parts as well as in the technical parts. This allows both parties to obtain a contract which is most fit for its use, minimising over- and under-provision. The user can expect lower prices due a more competitive market, and the provider can adapt the offer to the market conditions. The accepted (pre)standard now is WS-Agreement (version of March 2007), which is (March 2009) in its last steps to become a full standard. Several implementations of this protocol have been developed in the last two years, and are available on the GRAAP-WG website (most are open-source). The technical media to perform bargaining, the WS-AgreementNegotiation protocol, is not yet stable. It includes an extra getQuotes operation, which is implemented in the BEinGRID component.

The SLA Optimization matches the information offered in SLAs to the available resources. This improves the provider's scheduling strategy, allowing the provider to improve the utilization of its resources. It also allows implementation of the business rules which govern the allocation of resources based on the return value of the incoming SLA requests. Most schedulers are designed to optimize the resource usage based on the incoming resource requests, but very few take into account the business value of the request (the SLA describes the service request and its business context). This addition is a clear innovation, allowing the selection of jobs based on their value, pushing one step closer to an open and competitive service marketplace.

SLA Evaluation compares sensors output from monitoring tools to the SLA objectives, and raises alarms when thresholds are passed. The provider, having detailed information of its resource status, can act proactively to address failures, thus lowering the penalties incurred. The client can also receive these notifications, and can reallocate the job, augmenting the reactivity to failures. The accounting information is clearer, and claims for compensations are easier to make based on flagged violations. This system raises the confidence in the provider through more transparency. Monitoring tools are easily installable on a production system, like Nagios or Ganglia. But the evaluation of the SLA requires modelling business rules, then evaluating the functions yielding the violation status. Some previous results stemmed from the TrustCoM European project [8]. The evaluation functionality offers another management tool to discover the status of the service contracted.

SLA Accounting should be based on one of the existing charging schemes. As the metrics included in the SLA that deal with various and heterogeneous resources, the selection and adoption of the suitable charging scheme for each service executing environment allows charging the service based on its real execution cost. This connects resource usage more closely to retribution and penalties. The accounting concept has been studied extremely deeply, and many solutions exist. But adapting these to dynamic service provisioning with dynamic service conditions, as in the case of SLAs, requires some adaptation. One existing automatic solution should be selected and adapted to accept the Guarantees expressed in the SLAs, and seamlessly integrate SLAs as another means of describing service offerings.

In the current industrial setups, SLAs are long lasting contracts with detailed clauses, evaluated on long time-periods, and which are not automated (requiring legal signatures). This is not compatible with the current developments of SaaS, which allows its adopters to leave the silo and lock-in models for a more open market. Dynamic SLAs, signed when the service is needed, and easily finalized, will provide a more competitive market, where the user is more confident in his/her liberty of choice, while providing a more rational model for comparisons.

9.1.3 License Management

In order to foster the adoption of Grid technology in European business and society BEinGRID has gathered the requirements for a commercial Grid environment from a first wave of 18 business experiments. One of the key elements derived from this elicitation of requirements is support for commercial applications from independent software vendors (ISV) in grid environments. Especially small and medium enterprises (SME) from the engineering community stand to profit from pay-per-use HPC Grid scenarios (Utility Computing). Very few enterprises however maintain their own simulation applications. Instead—in contrast to academic institutions—commercial applications from independent software vendors are commonly used with an associated client-server based licensing. The authorization of these client-server based license mechanisms relies on an IP-centric scheme—a client within a specific range of IP-addresses is allowed to access the license server.

Due to this IP-centric authorization, arbitrary users of any shared (Grid/Cloud) resource may access an exposed license server, irrespective of whether or not they are authorized to do so. Secure and authorized access to a local or remote license server in Grid or Cloud environments therefore has not been possible so far. The use of commercial ISV applications in these environments therefore was not possible either.

This readily implies that the vast majority of users from industry have not been able to use ISV applications in grid environments. The License Management Architecture presented here is the first complete License Management Solution for Grids or Clouds and potentially increases the market size in the area of on-demand Grid/Cloud computing by industry by a large factor. The solution is generic, independent of specific middlewares and applicable to a large number of scenarios. The solution also features a cost-unit based accounting. In combination with secure access to the license server the developed License Management Architecture also supports the transition towards a true pay-per-use business model for licenses.

9.1.4 Data Management

Data management is important in a wide variety of business situations. For example the use of complex supply chains is increasing and in order to effectively plan production, companies have to access information from many suppliers. This data can be heterogeneous and held in many different locations which can cause problems in accessing, translating and using it. Solving these new Data Management problems is vital for improving the efficiency of modern organisations. It is also an opportunity to enhance existing Grid solutions for managing data to meet these challenging business requirements.

The first problem identified during the course of the project was how to easily access data remotely. Middleware such as OGSA-DAI already provides remote access to data (as well as a powerful and flexible workflow mechanism to manipulate the data) but applying the existing technology solution in new areas means engaging with users who may be unfamiliar with the technology and with Grid middleware in general. In order to make the business benefits of middleware like OGSA-DAI available to as many as possible it is important to lower the barriers to adoption. The Data Source Publisher is a mechanism to achieve this and it allows for the rapid and simple deployment of OGSA-DAI and the other Grid software it depends upon. It allows an organisation to quickly publish data and makes it easier for them to do so. This has a number of business benefits:

- New opportunities for collaboration with different organisations
- Better results from access to more sources of information
- Reduced costs due to better integration of data across multiple sites
- Larger markets for products and services.

As soon as you start to increase the possible sources of information available to an organisation you risk introducing a new set of problems. One of the other important requirements that were identified during the analysis of the Business Experiments was a need to access heterogeneous data. For example the organisations in a B2B network may all be using different software to manage orders and invoices. These legacy applications will use different databases with different schema. Again the idea of simplicity is very important. It is much easier to reason about data if it all looks the same. An important capability that we identified was to be able to homogenise data sources. This allows the differences in data to be hidden behind an abstraction layer. An example of this capability in action is the SQL Views component developed by the OGSA-DAI team in conjunction with BEinGRID.

In relational databases, SQL views are a common way of associating the results of an SQL query with a name. This name can be used as a short-hard to represent complex query results. It allows a table to be presented as having a different schema. OGSA-DAI's SQL views component provides an implementation of SQL views in OGSA-DAI. The SQL views component defines a mapping from view names to SQL queries representing these views.

Again using these techniques brings a number of benefits to the business willing to invest in the technology:

- Reduced development costs because applications can be simpler
- New opportunities for collaboration with different organizations
- Better results from access to more sources of information.

A reality of the business world is that though it may be ideal to start from a blank sheet of paper and design a totally new system, this is often simply impossible. Grid solutions must be able to interact with existing systems to gain a foothold. In light of this, another important requirement to emerge from the business experiments is a need to respond to changes of data. If you can meet this requirement you gain a number of benefits:

- Reduced development costs as existing applications can continue to use the database as before
- New markets as software can be developed to work alongside existing software or to extend competitors applications.

The OGSA-DAI Trigger component which has been developed as part of the BEinGRID project is a possible solution to these problems which also adds a new dimension to the existing Grid middleware. It provides a mechanism to execute any OGSA-DAI workflow in response to changes of data in a relational database.

These key problems and their solutions provide a very interesting area for continued research. The solutions identified and developed throughout the BEinGRID project already shown the real benefits to business that developments in this area can produce.

9.1.5 Security

The activities of the General Security (GS) area have led to the identification of Technical Requirements, Common Capabilities, Design Patterns and Software components to address issues of trust & security of users & applications in a distributed environment, typically regarding the privacy, confidentiality, and integrity of message exchanges between different users & services.

Details on the activities and results of the General Security area can be found in Chap. 4. We quickly summarize hereafter the main challenges & benefits encountered in the area of general security.

The key challenges come from the evolution of the way businesses interact nowadays: the work environment has become more pervasive with a mobile workforce, outsourced data centres, different engagements with customers and distributed sites. Systems are no longer monolithic: they integrate different services and clients from potentially many partners; each one with different security rules, identity stores, interfaces and regulations. Message exchanges no longer take place within the enterprise but across uncontrolled public networks. This stresses the need to *secure end-to-end transactions* between business partners and the customer. Companies will have to comply with their own directives and regulations as well as their partner organisations' rules and legal constraints: *compliance* must be monitored. In order to enable rich & flexible scenarios, the security mechanisms put in place must support, not hinder them and must *be flexible and adaptive*. Different enterprises, services and customers imply *multiple authorities* and complex relationships regarding the ownership of resources and information across different business contexts and organisational borders. Security policies must be issued by multiple administrators and enforced over a common infrastructure. There is also a need for well-orchestrated, end-to-end Operations management that provides controlled visibility, governance of network and IT state, timely assessment of the impact of security policy violations and the availability of resources. Hence, there is an increasing interest in *security observers & monitors*.

One can also refer to the challenges elicited in the Virtual Organisation thematic area (Chap. 3) to complete those already mentioned in the previous paragraph.

Five components have been developed by the General Security area over the course of the project to address these issues.

In particular, the Security Token Service (SOI-STS) (see Chap. 4) is an identity broker & federation manager that manages (a) an enterprise's participation in federation; (b) identity bridging between intra- and inter-enterprise identity technologies, claims, and authentication techniques; and (c) the lifecycle of identities and security attributes of users and services within that given enterprise. For more information on this capability please refer to Chap. 4. By federating identity brokers, a group of collaborators may create manageable circles of trust, each of them corresponding to a structurally rich trust network. The SOI-STS enables multiple administrators to control their own view of a circle-of-trust and authorized users & services. By issuing identity tokens, the SOI-STS also provides cryptographic material that can be used in secure e2e communications.

The Secure Messaging Gateway (SOI-SMG) (see Chap. 4) is a policy enforcement point and an XML Security Gateway which is software that enforces XML and Web service security policies. The SOI-SMG allows the enforcement of message and service-level policies with little or no programming. Combined with the SOI-STS or on its own, the SOI-SMG is able to analyse message flows, encrypt/decrypt, sign/validate signatures and again guarantee secure enterprise to enterprise communication. Because it is policy-based and its policy location mechanism is flexible, the SOI-SMG can allow for rich and diverse scenarios and deployments. Commercial alternatives also come with rich monitoring tools. Some of the key benefits of the SOI-SMG are that it decreases cycle time by removing security development burden from developers and coherently applying security policies across an entire enterprise. The SOI-SMG is also dynamically updatable, enabling definition of scenarios that evolve over time.

The Authorization Service (c) (see Chap. 4) is a policy-based authorization service which takes in access control requests, evaluates them against internal policies, and returns its decision to the requestor, typically the SOI-SMG. The SOI-AuthZ-PDP grants distributed access control and combines several access control models (attribute-based, role-based, rule-based) to produce an authorization framework suitable for highly distributed, dynamic environments. Because it uses the extended access control mark-up language (XACML), the SOI-AuthZ-PDP supports delegation which in turns enables a multiple administrative model where administrators from different realms can author access control policies. Its main benefit is the support of decentralised administration of access policies.

The Security Observer (SO) is a component that aims at monitoring security properties in a Grid environment and notifying subscribed entities when something wrong has been detected on these properties. As many Grid resources are heterogeneous and deal with numerous different technologies, the associated security can become heavy to process and to maintain. In order to centralize monitoring of possible security breaches and to relieve Grid entities from security routines, the Security Observer monitors various properties and can notify any program through a standard publisher/subscriber model. The Security Observer brings a centralized and common point for security information all over the Grid environment.

The security components can be brought together in order to create a richer, finely adaptive solution where, from an operational perspective, the SOI-SMG acts as an integration node which delegates authentication requests to the SOI-STS, authorization requests to the SOI-AuthZ-PDP, and is coupled with the security observer to monitor a certain set of parameters relating to the state & health of the entire infrastructure. Brought together, these components deliver a sturdy foundation for end-to-end web services security and generally speaking SOA security.

Overall, the expected benefits fall into two categories. Firstly, the security capabilities aforementioned help in being 'right first time'. By this, we mean that it becomes simpler for administrators to define, apply, and monitor security mechanisms. In particular, it becomes possible to write and execute different policies for different collaborations and keep them segregated. Therefore, services can be exposed several times in different business contexts with different security requirements & mechanisms in place tailored to the customer's specific needs. Because all components are programmatically manageable and customizable for different contexts (segregation of policy execution), it is possible to differentiate policies & services used in different collaborations with different customers. This also means we can use multiple security providers and integrate with 3rd party security services. More importantly, from a 'right first time' perspective, we can also assess the correctness of security enforcement via the validation of the declarative policies used in the different security components (AuthZ, AuthC, . . .). Another consequence of policy-based security components is regulatory compliance: this is achieved via policy coordination and their ability to be audited. The second benefit category is that of 'cycle time': time-to-market is greatly reduced when using such an infrastructure. Using a common security infrastructure that is flexible, scalable and dynamic reduces security management overhead as well as integration timescales of value-adding security services. The latter can also be outsourced to specialized 3rd party services. This lets enterprises exploit economies of scale by reusing a common security infrastructure in different collaboration contexts that they may not even need to maintain let alone create themselves anymore.

Chapter 4 delves further into details on each of the security components and their benefits.

9.1.6 Portals

As Grid Computing advances and becomes more widely used in the commercial and industrial sectors, the need for Grid environments that support multi-user applications is growing. Grid portals enable collaborative environments aiming to provide simple and common Web interfaces to heterogeneous Grid resources and services. The needed portal functionality, as identified in the requirement documents of the BEinGRID Business Experiments (BEs) [3], ranges from the submission, monitoring and control of computational jobs and visualization of their results to database access, management of remote workspaces and tools for accounting and billing. The portal should enable complex collaborations among heterogeneous systems and simplify administration and execution of all these operations.

The most fundamental requirements in a business environment are those of security and identity management. The Portals Security and User Management common capabilities were respectively designed to address the BE needs in these two closely related areas. User Management handles user account data and access rights to content within the portal, while Portals Security allows the simplified usage of different third party security systems in a common, integrated manner and is responsible for user access to external services. Portals Security enables the innovative concept of Single Sign-Up and an improved Single Sign-On mechanism for automatic user registration and authentication in both the portal and underlying Grid middleware. From the business perspective, the simple integration of third party security systems promotes easy incorporation of legacy systems in the portal, thus allowing preservation of existing investments in technologies and knowledge. The automation of specific security related processes has a positive impact on the security itself, but also on the efficiency and the responsiveness of the system. The abstraction of administrators from certain procedures decreases the human error factor and its severe consequences. At the same time, administrator expertise requirements can be lowered, which leads to reduction of administrative expenses. New users can be granted access to the system rapidly, which results in increased productivity and reduced cycle time, but also denotes end user confidence to the system.

Managing computational jobs via the portal was another important BE requirement. Job Submission Monitoring and Control (JSMC) makes it considerably easier for commercial users to use the power of the underlying computational Grid for justifying their needs, allowing businesses to expose a number of heterogeneous services through a common user friendly interface. Compatibility with existing systems allows reduced integration costs and preservation of investments in technologies and knowledge. New user training is also limited to just one user interface, regardless of the application or Service Provider (Grid platform) they need to use, thus minimising the costs related to the transition to a new portal environment for existing businesses. JSMC has the additional ability to store and display history and important details of previously submitted jobs, which is very important in terms of financial management and control of all job transactions held through the portal. The job control functionality helps minimising the usage of computational resources, which is directly associated with minimising operational costs of a business.

File Management offers the ability to access repositories of different types in a uniform manner through the portal, enabling improved management of data assets and improved reallocation of data resources in a simplified way, which is its main innovation and business value. File Management makes it possible to use existing file repositories via a new, user friendly portal environment. Its user interface enables easy execution of advanced file management or file transfer operations by utilising graphical interaction methods the average commercial user is already familiar with. Because of this fact, the users do not need to be excessively trained before being able to use the full file management capabilities of the Grid portal.

All common capabilities in the area of Grid portals aim at hiding complex operations and the complicated architecture of the Grid from novice users, and simplifying the procedures carried out through the portal by providing user friendly environments with improved usability and a familiar look and feel. These improvements contribute to increasing customer experience, reducing user frustration and increasing their trust in the system, and ultimately lead to increased productivity and profitability.

9.2 Analysis of Technical Requirements

In the technical area, the BEinGRID project has achieved several results in terms of common technical requirements (CTR), common capabilities (CC), design patterns and software components. The number of results are quite impressive (i.e. 39 CTR elicited, 36 CC defined, 32 Design Patterns produced and several Software Components developed or under development).

Effort has been spent also on analysing the dependencies between and across the achieved results. Without going into the technical details, it is worth analysing the correlations of these two results.

The diagram (Fig. 9.1) shows the CTR to CTR correlations. Here, positive correlation values are displayed as orange, from dark to light orange; values around zero are shown as yellow and light green, while negative values are coloured with shades of green. The legend shows the exact values for each colouring.

Generally, positive values indicate that the requirements occur together, negative values mean that requirements are more likely to be mutually exclusive and values of mean that there is no clear relationship between the requirements.

The high values around the diagonal of the picture shows that there is a strong correlation between requirements of each thematic area. The above correlation matrix has been refined for the developed components resulting in the diagram (see Fig. 9.2):

The idea behind the above diagram confirms the main trends that have already been identified by the Requirement to Requirement analysis: there is a strong correlation along the diagonal and some minor correlations in the independent thematic areas.

On the basis of the above results we can argue that:

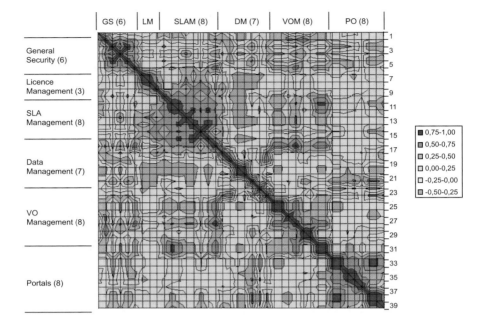

Fig. 9.1 CTR to CTR correlation

- There is a very strong correlation among results achieved inside each thematic area (the diagonal).
- Additionally the following technical areas show correlations:
  - GS with SLA, VOM and PO
  - LM with SLA and PO
  - SLA with GS, LM and (partially) VOM and PO
  - VOM with GS and (partially) SLA
  - PO with GS, LM and (partially) SLA and vice versa.
- Data Management results appear to be uncorrelated, or negatively correlated with most other technical areas. This may indicate a number of things.
  - The Business Experiments (BEs) with a particular focus on Data Management did not have time to explore other areas during the course of the experiments. For example the BE FilmGrid, which made extensive use of Data Management concepts identified the need for security but did not have time to pursue it fully during their experiment.
  - Business Experiments had existing results which they wanted to expand rather than a totally new problem to solve. For example GRID2(B2B) had an existing B2B platform and through the targeted use of Data Management components they were able to expand their existing solution which already handled the relevant other concepts identified by different technical areas.
  - Data Management is a core set of capabilities which underpin many different scenarios. In the initial analysis of business experiments, 9 of the 18 Business

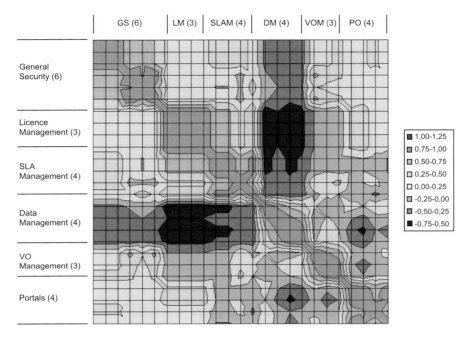

Fig. 9.2 Component to Component correlation

Experiments were identified as having interesting Data Management problems or requirements. Whereas requirements on areas like Virtual Organisations and SLAs may depend on the business model and logic, Data Management and also General Security underpin a wide variety of different, fundamental problems. A diverse set of scenarios can be enriched by the Data Management capabilities as demonstrated by the following examples:

* A Grid Architecture for Distributed Data [4]
* Portal Based Access to Grid Resources [5].

– Remarkably there is a negative correlation between Data Management and License Management.

License Management requirements predominantly appear in the area of HPC and Utility Computing. These scenarios in turn usually are associated with static collaborations and a vertical integration. In these scenarios Data Management—apart from non-functional requirements like network bandwidth—plays a minor role. On the other hand: If the scenarios are embedded in a larger context, Data Management requirements are likely to appear together with e.g. Roles and Rights Management. The negative correlation thus seems to be more related to the fact that the BEs were trying to avoid an overly complex setup.

The analysis presented above clearly show that the results achieved are correlated but do not give an indication on how to use collectively the results. This is also motivated by the fact that this project does not produce a common architecture but enables to build service oriented grid architectures based on the results of the technical activities (e.g. capabilities, design patterns, software components).

9.3 Overview of Common Technical Capabilities

As previously stated, the analysis done by the technical area of the BEinGRID project has shown that there are correlations among the technical results achieved and some key dependencies have already been considered in the development of components, such as the VO Set-up component presenting integration between components of the VO and GS thematic areas. But it still lacks an overall picture.

A first attempt to present an overall picture of the capabilities is graphically shown in Fig. 9.3. The figure groups all the Common Capabilities (CCs) into three different sets of general purpose capabilities to build a domain-neutral Grid based Service Oriented Architecture.

The sets are described in Table 9.1. See also the technical documentation at the Technical Solutions part of IT-tude.com [6].

It is clear that the view proposed in the Fig. 9.3 is additional and complementary to the basic thematic area oriented views coming from the BEinGRID approach. In fact, the CCs have originally been defined by taking into account the requirements specific to the six thematic areas.

So, while it is true that CC are still grouped per thematic area (this is shown with different colours in the picture), the classification proposed de-contextualises the CC from their thematic areas and re-groups them in a more generic approach.

Furthermore, it is difficult to find the 100% collocation of the CC into generic sets without loosing intuitiveness and clarity in the overall picture.

For example, some CC should be categorised into more sets. For instance, in addition to their collocation the PO-CC should be categorised also in the Visualization set. All PO-CC, in fact, are contextualised with respect to portals need and requirements. In some cases, we have considered predominant the functionality behind the portal (e.g. the PO-CC4 for file management) while in other cases the visualization aspect (e.g. the PO-CC3 for accounting).

Moreover, there are cases of CC providing similar functionalities, like DM-CC2 and PO-CC5, which both can access databases. The reason behind this is the CC have been defined on the basis of CTR elicited from different business cases. Some of the cases analysed heavily exploit portal technology and therefore need integration of functionalities inside portals.

An imperfect collocation of CC in generic sets as well as similarities among CCs is to be expected in initiatives of the size and complexity of BEinGRID where different groups have to work in parallel and meet tight timescales. However the governance process put in place to coordinate activities and ensure convergence of technical innovation, as described in Chap. 2 of this book, has provided for converging architectures that can be further adapted to produce combined products by

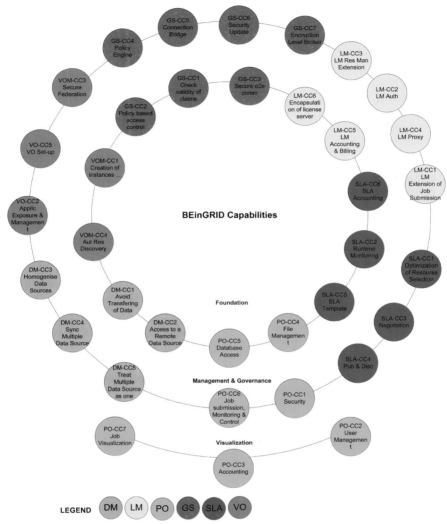

Fig. 9.3 The BEinGRID Common Capabilities

eliminating functionality overlaps when needed, and an interoperable implementation baseline, if capability implementations are used together as a loosely coupled service bundle. Hence, there is a strong foundation for providing an answer to the question: *how the technical results fit together to solve a complex problem*? The next section is devoted to describe this.

Table 9.1 The BEinGRID Common Capabilities

Foundation	CC that are considered fundamental to provide general purpose functionalities of a Grid based SOA

Among these we include capabilities to:

- Create service instances on a distributed environment (VOM-CC1)
- Access distributed data sources (DM-CC2)
- Allow creation of a secure context for policy based access control and e2e communication (GS-CC1, GS-CC2, GS-CC3)
- Encapsulate as web services license management servers for integration in SOA (LM-CC6)
- Allow run-time monitoring & accounting of the resources (SLA-CC2, SLA-CC6)
- Discover resources (VOM-CC4)
- Allow file management (PO-CC4)

Management & Governance	CC that are considered useful to provide general purpose functionalities for management of distributed resources, services and applications as well as for governance of a Grid based SOA

This set includes capabilities for resource, service and application management:

- Treating multiple data sources as one (DM-CC5)
- Optimise the resource selection (SLA-CC1)
- Allow secure exposure and management of services and applications (VOM-CC2)
- Publish and discover services (SLA-CC4)
- Allow job submission, monitoring and control (PO-CC6)
- And related capabilities with respect to license management:
  - LM related Extension of Job Description & Submission (LM-CC1)
  - LM related Resource Management Extension (LM-CC3)
  - License Proxy (LM-CC4)

Capabilities for governance:

- Updating policy and rules:
  - Policy Engine (GS-CC4)
  - Connection Bridge (GS-CC5)
  - Security Update (GS-CC6)
  - Encryption level Broker (GS-CC7)
- Automatically configure the distributed environment
  - Secure Federation (VOM-CC3)
  - VO Set-up (VOM-CC5)

For sake of clarity, Fig. 9.3 does not divide the CC between management and governance capabilities

Visualization	CC allowing visualization of information and knowledge

This set includes:

- Job information visualization (PO-CC7)
- Accounting info on the usage of resources and services (PO-CC3)
- Information on user's (PO-CC2)

9.4 Integration Scenarios

The focus of this section is to stress the "plug-n-play" approach allowed by the BEinGRID Common Capabilities, validating it through integration scenarios. Two business scenarios have been developed, involving technical results coming from the different technical areas of BEinGRID. Section 9.4.1 presents an example integration of VO, security and SLA results. Section 9.4.2 shows the integration of portals, license and SLA management. Both sections focus on providing a realistic situation, present how the Common Capabilities address the challenges of the scenario, then present the business point of view, and finally offer a design based on the software components stemming from the various technical areas.

9.4.1 The Federated ASP Scenario

The federated Application Service Provider (ASP) scenario can be seen as the collaboration of several Service Providers (SPs), providing services that can be combined into applications addressing a customer need. This could not be achieved individually. In this model a new service offering provided by an ASP is potentially built from services offered from many different organisations.

This model exploits the SaaS paradigm combined with the increasingly popular web 2.0 mash-up approaches. The idea is that one can increase service offerings by taking atomic services from one or more providers and offer a new service that builds on top of the aggregation of these atomic services.

The model also clearly separates non-functional requirements (e.g. security, QoS, ...) from functional requirements and business needs. This is critical to achieve flexibility and dynamic service composition.

This scenario is presented to allow the reader an understanding of how technical results of the VO, Security and SLA areas can be used collectively to address the main issues of the scenario.

9.4.1.1 Description of the Scenario

ImaginarySalesForce.com is a company which specialises in offering SaaS solutions to its customers. It plans to extend its current business model by offering complete domain-specific solutions based on new business opportunities.

ImaginarySalesForce.com neither has the necessary resources, nor the interest in developing all these domain-specific applications. It therefore plans to integrate the existing services provided by different organisations and decides to offer a new delivery channel to the small and medium sized software firms that currently supply specialized applications in the selected sectors.

In order to provide its customers with the required Quality of Service (QoS) ImaginarySalesForce.com will establish agreements with Infrastructure providers

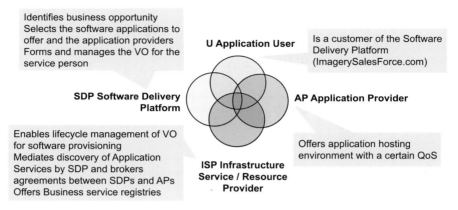

Identifies business opportunity
Selects the software applications to
offer and the application providers
Forms and manages the VO for the
service person

U Application User

Is a customer of the Software
Delivery Platform
(ImagerySalesForce.com)

**SDP Software Delivery
Platform**

AP Application Provider

Enables lifecycle management of VO
for software provisioning
Mediates discovery of Application
Services by SDP and brokers
agreements between SDPs and APs
Offers Business service registries

Offers application hosting
environment with a certain QoS

**ISP Infrastructure
Service / Resource
Provider**

Fig. 9.4 Federated ASP Actors

to host the necessary Infrastructure services, such as those for SLA Management, Security and VO Management.

The process of integrating new Application Providers (APs), Infrastructure Service Providers (ISPs) and Resource Providers (RPs) into ImaginarySalesForce.com will be conducted off-line given the necessary legal agreements for establishing the collaboration.

When ImaginarySalesForce.com identifies a business opportunity and adds a new application to its service portfolios, and once both firms have reached a convenient legal agreement, the process of VO formation starts. Delivering a new application also requires the establishment of agreements among the Software Delivery Platform (SDP) and Infrastructure (ISP) and Resource Providers (RP). During the VO formation phase the services belonging to the AP, ISP, and RP are registered in the SDP Registries. These services are then available to be used by the SDP when a customer of ImaginarySalesForce.com demands the execution of an Application.

9.4.1.2 Main Challenges of the Scenario: How Complementary CC Can Help?

The purpose of this section is to describe how the BEinGRID Common Capabilities can be used to solve common problems associated with the Federated ASP scenario. Indeed, the most challenging problem of this scenario is the governance and management of the collaborations among the actors involved.

Several challenges, in fact, are related to lifecycle management of the collaboration: partner identification and selection, agreement negotiation, establishment of secure federation among different administrative domains, identity brokerage, set-up of the distributed infrastructure underpinning a VO, service creation, secure exposure and management, partner and services monitoring and replacement, update of the infrastructure when context changes.

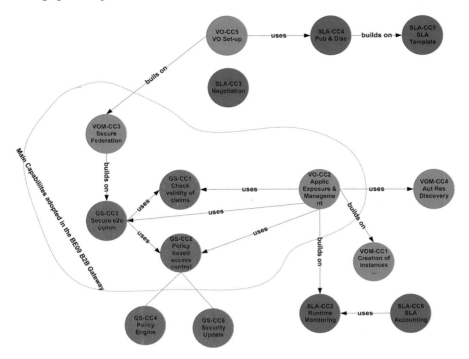

Fig. 9.5 CC to address the VO Governance and lifecycle management problem

In the following we are going to describe how the CC can be collectively adopted to address the above mentioned challenges. Figure 9.5 gives a graphical view of how the CC can be used for this.

In the picture, we show two types of relationships among CC: "builds on" and "uses". The "builds on" relationship indicates that a capability requires another capability to work properly while the "uses" relationship is a very generic relationship indicating that a capability may be combined or composed with others to provide a new functionality. For instance, in the figure "VOM-CC5 builds on VOM-CC3" means that the VO set-up capability (VOM-CC5) requires the Secure federation (VOM-CC3) component, while "VO-CC5 uses SLA-CC4" means that we can combine the VO Set-up capability with the publication and discovery of services (SLA-CC4) on the basis of SLA templates in order to offer a mechanism to identify partners on the basis of the services and associated SLA that they can offer. It is no surprise that "builds on" relationships are common among CC of a single thematic area while "uses" relationship is more common across thematic areas.

The picture also shows how some of the CCs have already been combined in the B2B Gateway product of the business case: "A Virtual Hosting Environment for an online distributed gaming application" [1].

In the following we explain how the CC can address the VO governance and lifecycle management and, in particular, how the CC can help support the Partner

Identification & Selection phase, Agreement Negotiation, Secure Federation, the operational and the evolution phase of a VO.

- Partner identification and selection
  The VO Set-Up (VOM-CC5) can be used in combination with the SLA Publication and Discovery (SLA-CC4) to allow the identification of partners in a VO on the basis of the capabilities they can provide and the associated SLA. The SLA-CC4 builds on top of the SLA template capability (SLA-CC5) to allow the publication and discovery of services on the basis of SLA templates.
- Agreement Negotiation
  Once partners have been identified and have a previous framework agreement, the negotiation of an SLA among the parties for the specific (aggregated) service at hand can be agreed using the SLA negotiations capability (SLA-CC3).
- Secure Federation
  The VO-CC5 builds on the VO Secure Federation capability (VO-CC3) to create a circle of trust among VO partners, defining to what level they trust each other. The VO-CC3, in turn, builds on the General Security CC to allow secure e2e communication (GS-CC3), to prove the validity of claims made by different parties (GS-CC1) and to allow secure policy-based access control to the resources of the VO (GS-CC2).
- Support of the Operational phase
  During the operational phase, the business process of the VO is executed and the performance is monitored. Service instances have to be created, exposed to the VO partners and managed by the VO partners as well. This is done via the Application Exposure and Management capability (VOM-CC2) that builds on the Creation of instances capability (VOM-CC1) to create services instances on the distributed infrastructure and on the runtime monitoring capability (SLA-CC2) to monitor the performance of service execution (via the associated SLA).

  The VOM-CC2 can be used in combination with the Automatic Resource discovery capability (VOM-CC4) to select the resources where services/tasks have to be executed, and with the GS-CC1, GS-CC2 and GS-CC3 to protect service/resource access and ensure secure e2e communication.

  Service execution is monitored using the SLA-CC2 and accounting information is gathered via the SLA-CC6.
- Support to the Evolution phase

SLA-CC2 allows monitoring of SLA. In case of violation of the service level agreement, new partners and services can be identified to replace the underperforming ones.

9.4.1.3 Business Benefits of the Scenario

There are several positive business aspects for the ASP implementing this scenario.

The Federated ASP scenario basically is an advanced ICT environment for: (i) integrating business services across enterprise boundaries, and (ii) virtualizing the (cross-organisation) ICT environment where these services operate.

Among the business benefits we recall, of course, the possibility to gain revenue from provision of applications, agility in providing the business (e.g. on-demand creation of virtual organisation to establish the business), cost savings of hosting all the services required for an application and management of infrastructure, reduction of the total cost of ownership by outsourcing parts of the value chain and, lastly, easy and transparent use of the Service Oriented Grid infrastructure, as well as enabler of new SaaS provisioning models.

9.4.1.4 Overview of the BEinGRID Components Used in the Scenario

This section presents an overview of the BEinGRID components that can be used in the scenario and show how they can interact together.

The components required for the provision to ImaginarySalesForce.com users with third party applications are presented in three phases: secure access, VO formation, and VO operation. The final phase, decommission, is not addressed here for clarity (it would involve final accounting, destroying the services, and revoking the VO).

1. *Phase1—Portal secure access and user management*: ImaginarySalesForce needs to manage its customer's identities efficiently. While integrating together different enterprises and various providers, ImaginarySalesForce needs to provide a mechanism to bridge and broker identities. In particular, once inside the collaboration, a particular user's identity within its own enterprise no longer makes sense. Therefore, there is a need to translate this identity and additional security attributes into a collaboration-wide virtual identity. To achieve this, we can use GS's Security Token Service (SOI-STS) which brokers identities and manages the release of security attributes depending on the context of the collaboration, the user, and the service targeted. The SOI-STS needs to connect to an internal enterprise user directory, e.g. an LDAP, MSAD. In this case, we envisage that we can integrate the SOI-STS with Portals' User Management component. The latter covers the need for Web interfaces managing portal user accounts, and their access to content or resources. This also includes the need for users to manage their own personal information and view information of other users, if authorised. This integration is further elicited in a whitepaper at [6].

2. *Phase2—VO Formation*: Once a user logs into in the ImaginarySalesForce.com portal and asks for the execution of an Application, the VO formation process starts:

   (a) All the capabilities provided by the ISV and ASP have to be exposed through a single access point, the ImaginarySalesForce.com Application Gateway & Secure Messaging Gateway. The Application Virtualization component addresses a common issue in SOA and Grid environments regarding the need to put together heterogeneous and distributed resources coming from different providers. It allows resources to be seen as a virtual common resource and provides the capacity of managing them as such. This component can be used as the Application Gateway for ImaginarySalesForce.com, which offers all the services through this single access point [Application Virtualization Component].

(b) When the user wants to use the semantically-enriched descriptions to find the ideal Application, the SDP checks its registries and searches for the candidate Application Services that meets a certain set of functional criteria. The Resource Discovery component improves the functionality of GLOBUS MDS (Monitoring and Discovery System), by providing an ontology-based interface, in order to locate resources, particularly computing resources, that satisfy a given requirement. This component allows the SDP to select the Application Services that meets a certain functional criteria (the decision of who decides on the criteria, and how the selection is done, can be responsibility of the SDP for ease-of-use, or of the user for finer control).

(c) Once the Application Services have been identified, the SDP has to identify the available Infrastructure Service Providers and Resource Providers that can provide the Application Services and Infrastructure Services with the required QoS and security requirements. It is necessary to negotiate an SLA Agreement between the SDP and the ISVs in order to assure that ImaginarySmalesForce.com provides the final user with the appropriate QoS. Each provider exposes its capabilities by means of an SLA Template. The SLA Negotiator Component implements the interfaces suggested by the March 2007 WS-Agreement specification that provides a simple protocol for negotiating SLAs. This component allows the creation of an SLA which governs the QoS at which a service is offered to its user. This component is accessed by the SDP which will negotiate the contract in the context of the user's application. When the provider is identified and the contract is signed, the infrastructure services to run and manage the application on a distributed set of resources or endpoints must be created. The provider only tackles resources within its own control; the SDP is in charge of passing the hardware resources secured to the AP. The information related to the real service instances offered to the user is available at the end of the negotiation phase. The component should also rely on an internal component to optimise the resource allocation, i.e. deciding which resources will be provided during the negotiation, and which allocation should be made. This is an internal service of the provider aiming at optimizing the scheduling of the jobs/services/resources [Automatic Resource Discovery component, SLA Negotiation component].

(d) The creation of a Secure Federation is requested among all involved providers, so that all the parties that belong to the VO can join together and communicate securely in accordance with the trust relationships created. This setup step has first been used with the Base VO, which contains all the actors of the marketplace, allowing secure communication for information retrieval. During negotiation, the partners forming the Operational VO are selected, and the selected security policies are further refined, to allow the privilege increase of the actor who will be calling the services. The VO-Setup component has a key role in finalizing the Virtual Organization creation process. It accomplishes the basic steps that are

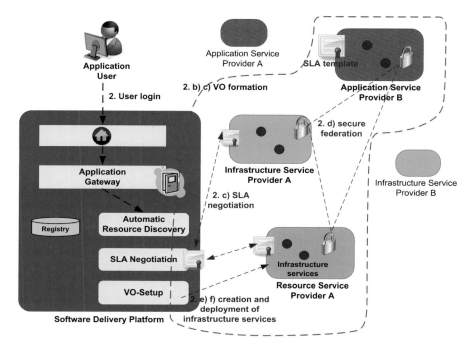

Fig. 9.6 Phase2—VO Formation

needed to set up a VO (after memberships, roles, agreements, and contracts have been negotiated and the VO policies have been defined). The main objective is thus formalizing the collaboration among partners. It gathers the different services that have been contracted and does the last operations needed for the use of the complete infrastructure by the SDP (VO Set-up component interacting with the GS components namely the SOI-STS for creation of secure federation). The interaction with the SOI-STS is a management one at this stage. In particular, with regards to the creation of secure federations, the SOI-STS, which acts as an identity broker as well as a circle-of-trust implementer, receives the relevant partners' business cards and adds them to a newly created local federation view. The SOI-STS will also maintain a list of services and users (along with their credentials and identity attributes) within that particular federation view. Other security services that are preselected at this stage include the SOI-AuthZ-PDP, an authorization services which controls access to resources in the collaboration, and the SOI-SMG, and the secure messaging gateway which enforces the exposure & security policies for each resource.

(e) When all the Application Service providers and Infrastructure Service providers are identified, it is necessary that all involved providers create the Infrastructure services to run and manage the application on a distributed set of resources or endpoints. [Application Virtualization component that further configures the SOI-{STS; AuthZ-PDP; SMG} and SLA monitoring and evaluation components.] The SOI-SMG and the SOI-AuthZ-PDP in particular will be instantiated within the particular federation created for the service instance being exposed. The security services ensure the segregation of policy execution thus giving a contextualized experience and exposure. Typically, each provider has its own SOI-{STS; AuthZ-PDP; SMG}. In particular, the SOI-STS will maintain that provider's view of the federation. Note however that each of these security services can be exposed as SaaS and consumed from the cloud.

(f) Afterwards it is necessary to perform configuration operations: configuration of the infrastructure, instantiation and orchestration of the Application service, assignment and set up of resources and activation of services, notification to the involved members and manifestation of the new VO (VO-Setup component). Configuration operations also include the definition and refinement (as previously) of security policies. These security Policies govern the behaviour of the identity broker (SOI-STS), secure messaging gateway (SOI-SMG) where the services are eventually exposed, and the authorization service (SOI-AuthZ-PDP). A coherent and real-time view of the system is necessary to make sure the services are adequately protected.

3. *Phase3—VO Operation*: Once the user is using the application, it is necessary to:

(i) Establish mechanisms to validate that the ISV and Resource Providers are providing the required QoS at run-time. On one hand, at the provider's side, mechanisms to monitor the service provision based on low-level monitoring data have to be established. On the other hand, at the client side, mechanisms are needed to evaluate the providers' gathered information to assure that the previously agreed SLAs are honoured, and the appropriate actions can be taken in the case of violation. This is done via the SLA Runtime Monitoring and Evaluation component.

(ii) Ensure the secure communication among participants. During the operational phase, secure communication is ensured by the triplet SOI-{STS; AuthZ-PDP; SMG} of the security component that should be deployed on each partner site. Because the SOI-SMG, SOI-AuthZ-PDP and SOI-STS are all built on top of common web service standards (WS-Security, XML Encryption, WS-Trust, . . .), they are easily integrated. It is worth remembering that the SOI-SMG focuses on XML security enforcement and is as such the central part of the integration between all three components. It delegates further security calls to the two remaining components. The SOI-AuthZ-PDP focuses on XACML-based access control, and the SOI-STS on identity brokerage. Typically, during an outbound call, the SOI-SMG at the client side would request a virtual identity token on behalf of the user from the client's SOI-STS. That token can carry identity attributes that can be used on the service-side by the service SOI-SMG, when it invokes the SOI-AuthZ-PDP for an access control decision. More information on this integration and set of interactions is provided in the section on General Security and in [2].

9.4.2 The Collaborative Engineering Scenario

This section is about collaborative engineering, where design and development of a complex product is done by different geographically distributed organizations or engineering teams that, to this purpose, share resources, information and knowledge. It involves technical results mainly coming from the Portals, License management and SLA areas. The scenario will demonstrate how these technologies can be used to build an infrastructure for collaborative engineering where software, engineering and compute cycle providers are involved. For clarity we reduced it to a unique service provider, but it is extensible to a more complete value added chain, where for example database access for product life cycle management will be included. Examples are given by the business experiments of the first phase of BEinGRID.

9.4.2.1 Description of the Scenario

This scenario depicts the development of a complex product whose parts are produced by different contractors, in turn employing consultants who evaluate the designed parts by the means of computational engineering. We see for example vehicle manufacturing or ship building as tangible instances of such a model.

Figure 9.7 depicts the actors of this scenario.

The topmost shape represents the prime contractor or controller of the project, which is the only instance with access to all parts of the final product. The second level consists of companies supplying only parts of the product, having limited access to the material and interfaces of their parts. To optimise the final design they employ consultants to analyse the product performance by the means of computer simulations. This simulation is carried out on the third party resources using software by independent software vendors. A different shape may not necessarily represent a different individual organisation, but likely teams of the same organisation or consortium.

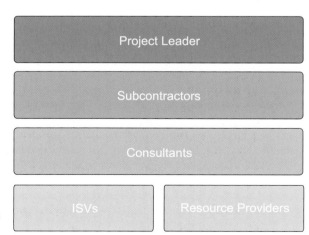

Fig. 9.7 Abstract collaborative engineering role schema

9.4.2.2 Main Challenges of the Scenario: How Complementary CC Can Help?

The purpose of this section is to describe how common capabilities can be used to collectively solve common problems associated with the Collaborative Engineering scenario.

Collaborative Engineering scenarios have several characteristics that can successfully be addressed with the set of BEinGRID CC. Typical features are high level of confidentiality of data, large data sets, high computational demand, rather fixed set of participants, and rather long lived collaborations

Figure 9.8 shows the flow of activities in the scenario assigned to the different roles. This flow of activities is then addressed via the BEinGRID common capabilities.

The user access and interaction is supposed to be conducted completely over a Portal.

The project leader is in charge of creating the project workspace to serve as the organizational scope of the project and managing the other users (PO-CC2). The different participants are associated with roles that allow them to execute different actions, both on the portal and on the resources of the VO, such as storage or computing resources. The various user roles allow access to different parts/pages of the portal and thus the underlying services and resources accessed through those portal pages. In a lower level, a user of the portal cannot access a service if no appropriate credential is associated with the user's portal identity (PO-CC1).

With the Portal File Management (PO-CC4) the participants are able to organise and exchange data securely (PO-CC1), and access the project's collaborative file repository, where input files, as well as job execution results, can be stored and shared between the different project contractors. The project leader can, for example, use the File Management component and collaborative file repository to provide the subcontractor with the required data about specifications and interfaces. After completing his/her work, the subcontractor will upload a virtual prototype of the desired product. By making this prototype or parts thereof available to the consultant, it will be possible to accomplish the pre-processing for the simulation.

The consultant will upload the prepared data again, which is submitted to a computational job from within the Portal, which will access this data from the remote site (DM-CC2). Through the Job Submission Monitoring & Control facilities, the consultant is able to overlook the execution (PO-CC6) of those computations and control them. For example, in case a wrong input file has been specified for a long-lasting simulation, the user can cancel it and start over again. The simulation results can be quickly visualised from within the portal (PO-CC7), to check if they are reasonable. In case the results are as expected, then the associated amount of data (usually a large one) can be downloaded and studied in detail at the user's local machine. This way, bandwidth usage can be reduced and processes can be sped up.

The ISV manages a License Server so that other project participants can use its software on the accessible Resources (LM-CC4). The License Management's Accounting facility (LM-CC5) can be used to monitor and bill the Resource Providers for the "on-demand" usage of the software. The Resource Providers take these costs into account when signing SLAs.

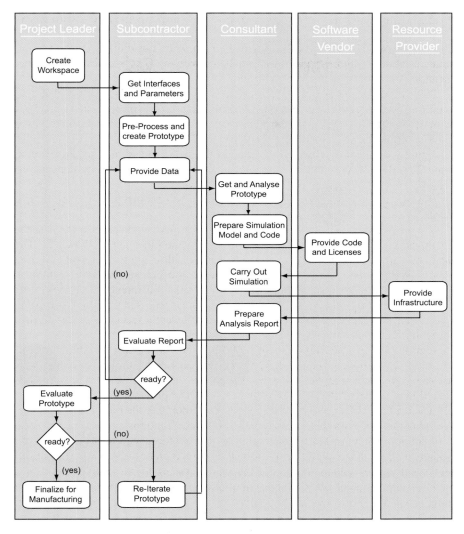

Fig. 9.8 Activity Diagram for Interactions between the different roles in the Scenario

For the interaction between Consultant, ISVs and Resource Providers, SLAs can guarantee a certain degree of quality of service or automatically negotiate terms to comply to fulfil a contract (SLA-CC2, SLA-CC3, SLA-CC6). The SLAs can be abstracted as another job resource, and as such propose the same access allowing to create, monitor, and account SLAs.

Information associated with SLA and LM accounting can be displayed through the portal (PO-CC3), so that the project leader has a view of the billing information and the resources in use in real-time.

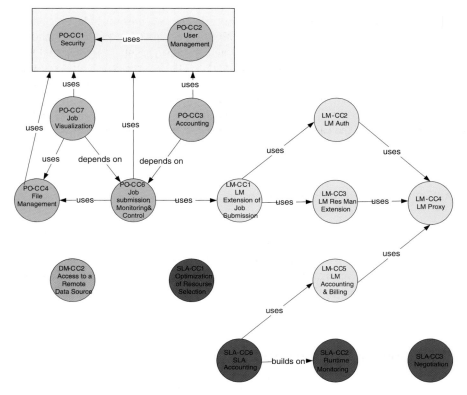

Fig. 9.9 Collective adoption of CC in the Collaborative Engineering Scenario

As the central system manages and transfers sensitive data, all communication must be adequately secured. This includes and goes beyond the interaction with the Portal, data transfer to the central repository, data transfer and access to and on the compute element and communication with the license management and SLA servers.

Due to the hierarchical nature of project participants, certain access policies need to be enforced during the process. As the data in the level of product parts is likely covered by non-disclosure agreements, it is essential to control who can access data of upper levels at what time. For example, the Subcontractors may only access the requirements and interface descriptions of the subsystems they need to connect to, whereas the Project Leader could perhaps be allowed to see all design drafts and analysis reports the subcontractors and consultants produce.

The following picture shows how BEinGRID CC can be collectively used to execute the activities presented in the sequence diagram of Fig. 9.8 (previously described).

Similarly to Fig. 9.5 (presented in the other scenario), the diagram (Fig. 9.9) shows two types of relationships among CC: "builds on" and "uses". The "builds on" relationship indicates that a capability requires another capability to work properly while the "uses" relationship is a very generic relationship indicating that a capability may be combined/composed with others to provide a new functionality.

In the picture, it is evidenced that user can access in a secure way the portal with the combination of PO-CC2 and PO-CC1. After the secure access, the PO-CC1 can be used in combination with the GS-CC3 to allow the distributed actors of the Collaborative Engineering scenario to interact in a secure way.

The diagram also shows the relationships between the PO-CC6 for Job submission and the CC of the License Management framework as well as the relationships between the PO-CC3 for accounting and the SLA CC.

9.4.2.3 Business Benefits of the Scenario

The business benefits in implementing this scenario are, of course, different for the specific actors involved. In general, we recall rapid product design and development, better cost estimation and control (by integrating cost model and design-to-cost processes), clear definitions of roles and tasks, reducing the risk of designing and development (assigning right tasks to the right people) and improved engineering analysis capabilities (by exploiting specialised tools and services).

9.4.2.4 Overview of the BEinGRID Components Used in the Scenario

Figure 9.10 shows an overview of the BEinGRID components applied in this Scenario. The Grid Portal represents the User Interface for all participants.

The Portals Security component provides the necessary facilities for user registration and authentication. It is used to register the users in the portal and the Grid middleware, to create the corresponding credentials, to sign and to store them in the component's internal credential repository. The component replaces the existing "new user sign-up" capabilities of the portal container and uses its own database to store new user account info. For each new portal user, a corresponding user account at the portal container is created and mapped with Portals Security. The same is done for every Grid middleware the user will have access to. This mechanism is called single sign-up. Each time a user logs in to the portal, related security information for that user (username, password, etc.) is retrieved and stored as session parameters or proxy certificates in the internal repository and used to also authenticate the user at the portal container. Then, e.g. during Job Submission and File Management operations, this information is passed to the external services, as appropriate according to the authentication method they utilize. This provides a single sign-on solution for the whole workflow. Portals Security also provides complete credential management functionality and the ability to use an external credential repository, like MyProxy, for optionally retrieving existing credentials, if so wished. User authorization is handled by the Role Based Access System of the portal container in a direct way, and indirectly by means of the Portals Security component.

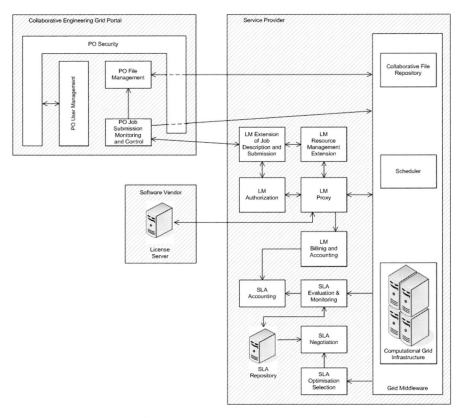

Fig. 9.10 Overview of Components used in the Collaborative Engineering Scenario

The User Management component handles user account information and enables users to edit their personal data or view detailed information of other users, if authorized. Administrators can use the User Management component to oversee and maintain user information. It handles arbitrary data about the users, like contact information, and helps organizing groups they are part of and resources that will be available to them by using the Role Based Access Control (RBAC) features of the portal container. It is closely related to the Portals Security component, because it enables mapping of user accounts in the portal, the portal container and the external services. The two components interoperate in a seamless manner to achieve the desired functionality.

The File Management component enables a Graphical User Interface that allows the upload, download and further management of files to a Collaborative Repository, which is internally accessed through the Grid Middleware, or to another external file repository, if so required. In addition, this component enables a personal storage space at the portal for each user. To leverage the Grid middleware authorization capabilities, the user must be authorized with the Grid Middleware. The component therefore uses the Portal Security component to authorize the user communication with the Repository.

Files that are stored in the Repository can further be used as input files for Jobs. The Job Submission Monitoring and Control (JSMC) component uses the functionality offered by the lower layers of the File Management component to access file repositories by defining staging-in and -out steps such as copying input files from the Collaborative Repository to the Compute Element prior to Job execution and copying result files after the computation.

The JSMC component allows a user to begin the execution of a computational Job by specifying the Job and its resource requirements. The user can for example select the Job's type and parameters, input files and the destination for output files. To influence where the Job will be executed, the user can further specify different performance criteria. For Jobs that need a provided license server, the LM Extension of Job Description and Submission adds the possibility to specify a license server and the corresponding required credential information to be used for license retrieval.

When the specification is complete, the user can initiate the submission of the Job. If a license will be required, the Job is also registered at the LM Authorization component to later on authorize the usage of licenses. A Job submitted from the portal will typically be handled by a scheduler at the Grid middleware. An interface to the scheduler could be proposed through the negotiation of an SLA using the SLA Negotiator component, or though some scheduler optimizer as the SLA Optimisation Selection component to choose the best available resource to submit to.

The LM Resource Management Extension component is used to monitor the availability of licenses and only permit the initiation of a new Job if enough licenses are available for its execution. The LM Proxy is potentially used to allow the connection to License Servers to check for license availability prior to Job submission and to retrieve the appropriate license when Job execution starts. Access to the licensing system is controlled by the LM Authorization component, and the usage is monitored by the LM Monitor component and accounted for through the LM Billing and Accounting component.

The SLA Evaluation and Monitoring component records the resource usage of the Job and supervises the adherence of the terms contained in the SLA that was provided to the SLA Repository at a prior phase through an SLA Negotiation procedure. SLA Accounting then processes the performance data from the SLA Evaluation and Monitoring and LM Billing and Accounting components to generate billing information.

References

1. Business Experiment 9 (BE9) Distributed Online Gaming, http://www.BEinGRID.eu/be9.html
2. T. Dimitrakos, D. Brossard, P. de Leusse, Securing business operations in an SOA. BT Technology Journal **26**(2) (2009)

3. Globus Toolkit Information Services: Monitoring & Discovery System (MDS), http://www.globus.org/toolkit/mds/
4. IT-tude.com Technical Solutions—A Grid Architecture for Distributed Data, http://www.it-tude.com/data_grid_arch.html
5. IT-tude.com Technical Solutions—Portal Based Access to Grid Resources, http://www.it-tude.com/portal_access_to_grid.html
6. IT-tude.com Technical Solutions, http://www.it-tude.com/technical.html
7. SPARQL Query Language for RDF, http://www.w3.org/TR/rdf-sparql-query/
8. TrustCoM European IST-FP6 project, http://www.eu-trustcom.com/

Printing and Binding: Stürtz GmbH, Würzburg